Mohammed Hossein Keshavarz, Thomas M. Klapötke
Energetic Compounds

Also of Interest

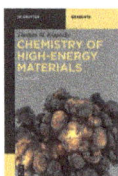

Chemistry of High Energy Materials
Klapötke, 2019
ISBN 978-3-11-062438-0, e-ISBN 978-3-11-062457-1

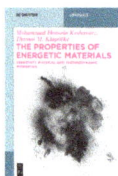

The Properties of Energetic Materials.
Sensitivity, Physical and Thermodynamic Properties
Keshavarz, Klapötke, 2017
ISBN 978-3-11-052187-0, e-ISBN 978-3-11-052188-7

High Explosives, Propellants and Pyrotechnics
Koch, 2020
ISBN 978-3-11-066052-4, e-ISBN 978-3-11-066056-2

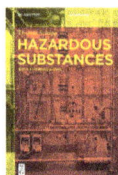

Hazardous Substances.
Risks and Regulations
Schupp, 2020
ISBN 978-3-11-061805-1, e-ISBN 978-3-11-061895-2

High Temperature Materials and Processes
Fukuyama, Hiroyuku (Editor-in-Chief)
ISSN 0334-6455, e-ISSN 2191-0324

Mohammed Hossein Keshavarz,
Thomas M. Klapötke

Energetic Compounds

Methods for Prediction of their Performance

2nd edition

DE GRUYTER

Authors

Prof. Dr. Mohammed Hossein Keshavarz
Malek-ashtar University of Technology
Department of Chemistry
83145 115 Shahin-shahr
Iran
keshavarz7@gmail.com

Prof. Dr. Thomas M. Klapötke
Ludwig-Maximilians-Universität
Department of Chemistry/
Energetic Materials Research
Butenandstr. 5-13
81377 Munich
Germany
tmk@cup.uni-muenchen.de

ISBN 978-3-11-067764-5
e-ISBN (PDF) 978-3-11-067765-2
e-ISBN (EPUB) 978-3-11-067775-1

Library of Congress Control Number: 2020934486

Bibliographic information published by the Deutsche Nationalbibliothek
The Deutsche Nationalbibliothek lists this publication in the Deutsche Nationalbibliografie;
detailed bibliographic data are available on the Internet at http://dnb.dnb.de.

© 2020 Walter de Gruyter GmbH, Berlin/Boston
Cover image: michalz86 / iStock / Getty Images Plus
Typesetting: VTeX UAB, Lithuania
Printing and binding: CPI books GmbH, Leck

www.degruyter.com

Preface to the first edition

Since the synthesis and development of new energetic materials require the identification of promising candidates for additional study and the elimination of unsuitable candidates from further consideration, it is important for engineers, scientists, and industry to be able to predict the performance of new compounds in order to reduce the costs associated with the synthesis, testing, and evaluation of these materials. Nowadays different approaches have been used to predict the performance of energetic compounds which have proven to be cost-effective, environmentally-friendly, and time-saving approaches. This book reviews different methods for the assessment of the performance of an energetic compound through its heat of detonation, detonation pressure, detonation velocity, detonation temperature, Gurney energy, and power (strength). It also focuses on the detonation pressure and detonation velocity of nonideal aluminized energetic compounds. Simple and reliable methods are demonstrated in detail which can be easily used for the design, synthesis, and development of novel energetic compounds.

Mohammad Hossein Keshavarz
Thomas M. Klapötke

https://doi.org/10.1515/9783110677652-201

Preface to the second edition

Everything said in the preface to the first edition still holds and essentially does not need any addition or correction. In this revised second edition, we have updated the manuscript and added some recent aspects of energetic materials:

(i) Some errors which unfortunately occurred in the first edition have been corrected and the references have also been updated where appropriate.

(ii) Recent works have been reviewed and discussed in each chapter. Moreover, new sections have also been inserted including application of laser techniques for assessment of the detonation performance, plate dent test, combined effects aluminized explosives, Gurney velocity of aluminized explosives and assessment of the detonation velocity of primary explosives.

(iii) A new chapter 7 about Underwater Detonation (Explosion) has been added.

Mohammad Hossein Keshavarz
Thomas M. Klapötke

https://doi.org/10.1515/9783110677652-202

Contents

List of symbols

a	Number of carbon atoms
A	Distance between the mark on the witness plate and center of the detonating fuse in the Dautriche method
A_{JWL}	Linear coefficient of JWL-EOS
AB	Parameter in Rothstein and Peterson's Method that is equal to 1 for aromatic compounds and is otherwise given the value 0
b	Number of hydrogen atoms
B_{JWL}	Linear coefficient of JWL-EOS
BKW-EOS	Becker–Kistiakowsky–Wilson equation of state
BKWC-EOS, BKWR-EOS and BKWS-EOS	Three different parameterizations of BKW-EOS
c	Number of nitrogen atoms
C_{JWL}	Linear coefficient of JWL-EOS
C_{polar}	Contribution of some specific polar or functional groups in aromatic and non-aromatic $C_aH_bN_cO_d$ energetic compounds in the prediction of the heat of detonation
C_{SFG}	Contribution of some specific functional groups in aromatic $C_aH_bN_cO_d$ energetic compounds in the prediction of the heat of detonation
C_{SSP}	Contribution of some specific structural parameters in non-aromatic $C_aH_bN_cO_d$ energetic compounds in the prediction of the heat of detonation
CHEETAH	Thermochemical computer code
C–J	Chapman–Jouguet
$\bar{C}_V(\text{detonation product})_j$	Molar heat capacity of jth product at constant volume
$\bar{C}_P(\text{detonation product})_j$	Molar heat capacity of jth product at constant pressure
d	Number of oxygen atoms
$D_{det}(\text{explosive charge})$	Detonation velocity of the explosive being tested using the Dautriche method
$D_{det}(\text{detonating fuse})$	Detonation velocity of the calibrated detonating fuse used in the Dautriche method
D_{det}	Detonation velocity
D_{metal}	Terminal metal velocity
$D_{det,max}$	Theoretical maximum density of the explosive
$D_{det,max}^{Dec}$	Correcting function for decreasing the value of $D_{det,max}$
$D_{det,max}^{Inc}$	Correcting function for increasing the value of $D_{det,max}$
DOE	Design of experiments
DSC	Differential scanning calorimetry
e	Number of fluorine atoms
E	Detonation energy per unit volume

https://doi.org/10.1515/9783110677652-203

E_b	Gas bubble energy per kg explosives at the measuring point
E_G	Specific energy or Gurney energy (J)
$\sqrt{2E_G}$	Gurney velocity or Gurney constant (m/s)
$(\sqrt{2E_G})_{H-K}$	Gurney velocity from Hardesty and Kennedy (H–K) method
$(\sqrt{2E_G})_{K-F}$	Gurney velocity from Kamlet and Finger (K–F) method
EOS	Equation of state
E_S	Shock wave energy
EXPLO5	Thermochemical computer code
f	Number of chlorine atoms
$\%f_{Trauzl,TNT}$	Relative power of an energetic compound with respect to TNT from the Trauzl lead block test
f^+_{Trauzl}	Correcting function for the adjustment of an underestimated value of $\%f_{Trauzl,TNT}$ obtained on the basis of the elemental composition
f^-_{Trauzl}	Correcting function for the adjustment of an overestimated value of $\%f_{Trauzl,TNT}$ obtained on the basis of the elemental composition
$f^-_{brisance,sand}$	Correcting function for the adjustment of an overestimated value of $\%f_{brisance,sand,TNT}$ obtained on the basis of the elemental composition
$\%f_{ballistic\ mortar,TNT}$	Relative power an energetic compound with respect to TNT from the ballistic mortar test
$\%f_{brisance,sand,TNT}$	Relative brisance of an energetic compound with respect to TNT for sand crushing test
$(\%f_{brisance,sand,TNT})_{aluminized\ explosive}$	Relative brisance of an energetic compound with respect to TNT of sand crushing test for aluminized explosives
$\%f_{brisance,plate\ dent,TNT}$	The relative brisance with respect to TNT for the plate dent test
f^+_{Trauzl}	Correcting function for the adjustment of an underestimated value of $\%f_{Trauzl,TNT}$ obtained on the basis of the elemental composition
$f^+_{brisance,sand}$	Correcting function for the adjustment of an underestimated value of $\%f_{brisance,sand,TNT}$ obtained on the basis of the elemental composition
g	Number of aluminum atoms
G	Parameter in Rothstein and Peterson's Method that is equal to 0.4 for liquid explosives and 0 for solid explosives
GA	Genetic algorithm
GIPF	General interaction properties function
H	Constant pressure enthalpy
h	Number of moles of ammonium nitrate in an explosive composition
$H_{poducts}$	Enthalpy of products
$H_{reactants}$	Enthalpy of reactants
H^θ	Enthalpy of the desired species under standard conditions (298.15 K and 0.1 MPa pressure)
$H(c)$	Enthalpy of the desired species in the condensed phase (solid or liquid)

$H^{\theta}(c)$	Enthalpy of the desired species in the condensed phase (solid or liquid) under standard conditions (298.15 K and 0.1 MPa pressure)
$H(g)$	Enthalpy of the desired gaseous species
HTPB	Hydroxy-terminated polybutadiene
I_{SP}	Specific impulse
ISPBKW	Computer code for calculation of the specific impulse using BKW-EOS
JCZS-EOS	Jacobs–Cowperthwaite–Zwisler equation of state
JCZS3-EOS	Jacobs–Cowperthwaite–Zwisler-3 equation of state
JWL-EOS	Jones–Wilkins–Lee equation of state
k_i	Molar co-volumes of the i-th gaseous product
K–J	Kamlet–Jacob
L	Length between the probes in the Dautriche method
LASEM	Laser-induced air shock from energetic materials
LIBS	Laser Induced Breakdown Spectroscopy
m	Mass of explosive charge
$\frac{m}{c}$	The ratio of the masses per unit length of the metal and the explosive
$M_{w\,gas}$	Average molecular weight of the gaseous products
MAPE	Mean absolute percentage error
$(\frac{n_{Al}}{n_O})_{LIBS}$	The Al/O intensity ratio determined by LIBS method
$(\frac{n_{Al}}{n_O})_{theory}$	The Al/O intensity ratio calculated by chemical composition
$n(g)$	Number of moles of gas involved
$n(HF)$	Number of hydrogen fluoride molecules that can possibly be formed from the available hydrogen
$n(B/F)$	Number of oxygen atoms in addition to those necessary to form CO_2 and H_2O, and/or the number of fluorine atoms in excess of those required to form HF
$n(C=O)$	Number of oxygen atoms doubly bonded directly to carbon
$n(C-O)$	Number of oxygen atoms singly bonded directly to carbon
n_{exp}	Number of moles of explosive
$n(NO_3)$	Number of nitrate groups present as a nitrate ester, or as a nitric acid salt such as hydrazine mononitrate
n_j	Number of moles of the jth detonation product
n'_{gas}	Number of moles of gaseous detonation products per gram of explosive
n_{mN}	Number of nitro groups attached to carbon in nitro compounds in which $a = 1$
n_N	Parameter that is equal to $0.5n_{NO_2} + 1.5$ where n_{NO_2} is the number of nitro groups attached to carbon in nitro compounds for the prediction of the maximum attainable detonation pressure in which $a = l$
n_{NH_x}	Number of $-NH_2$ and NH_4^+ moieties in the energetic compounds
n_{NR}	Number of $-N=N-$ groups or NH_4^+ cations in the explosive

$n_{-NRR'}$	Number of $-NH_2$, NH_4^+ or $\left[\begin{smallmatrix}-N\\ \ \ \ N\\ -N\end{smallmatrix}\right]$ groups
$n_{NR_1R_2}$	Number of $-NH_2$, NH_4^+ and five membered rings with three (or four) nitrogens in any explosive, as well as five (or six) membered rings in nitramine cages
n'_{Al}	Number of moles of aluminum atoms under certain conditions
n'_{NO_3salt}	Number of moles of nitrate salt
n_1^0	Parameter that equals 1.0 for energetic compounds that follow the condition $d > 3(a + b)$ and zero for other energetic compounds
p	Shock wave pressure
P	Pressure
PCA	Principle component analysis
$P_{cyc,nitramine}$	Correcting function for the prediction of the heat of detonation of cyclic nitramines
P_{det}	Detonation pressure
$P_{det,max}$	Detonation pressure at the maximum loading density or theoretical maximum density of an explosive
$P_{det,max,SSP}$	Parameter that is equal to 1.0 for explosives which contain N=N-, $-ONO_2$, NH_4^+ or $-N_3$ in the molecular structure for the prediction of the maximum attainable detonation pressure
P'_{in}	Correcting function that is specified for increasing the value of the maximum attainable detonation pressure on the basis of the elemental composition under certain conditions
P'_{de}	Correcting function that is specified for decreasing the value of the maximum attainable detonation pressure on the basis of the elemental composition under certain conditions
PLS-DA	Partial least squares discriminant analysis
Power Index $[H_2O(g)]$	Power index of an explosive if the water in the detonation products is in the gaseous state
Power Index $[H_2O(l)]$	Power index of an explosive if the water in the detonation products is in the liquid state
Q	Heat transfer
Q_{expl}	Heat of explosion
Q_{det}	Heat of detonation
Q'_{det}	Heat of detonation
$Q_{det}[H_2O(g)]$	Heat of detonation when H_2O in the detonation products is in the gas phase
$Q_{det}[H_2O(l)]$	Heat of detonation when H_2O in the detonation products is in the liquid state
$Q_{det}[H_2O(l)]_{aromatic}$	Heat of detonation of an aromatic high explosive when H_2O in the detonation products is in the liquid state

$Q_{det}[H_2O(l)]_{non\text{-}aromatic}$	Heat of detonation of a non-aromatic high explosive when H_2O in the detonation products is in the liquid state
$Q_{H_2O\text{-}CO_2}$	Heat of detonation based on the "H_2O–CO_2 arbitrary"
r	Distance between the pressure gauge and charge
R	Gas constant
R_1	Nonlinear coefficient of JWL-EOS
R_2	Nonlinear coefficient of JWL-EOS
$R-R_0$	Actual radial expansion in cylinder test
rms deviation	Root mean square of deviations
STP	Standard Temperature and Pressure
SVR	Support vector regression
T	Temperature
T_b	First pulsation period
T_i	Initial temperature
T_{max}	Maximum temperature
T_{det}	Detonation (explosion) temperature
$(T_{det})_{aromatic}$	Detonation temperature of an aromatic energetic compound
$(T_{det})_{non\text{-}aromatic}$	Detonation temperature of a non-aromatic energetic compound
TMD	Theoretical maximum density
V	Volume
V_0	Volume of undetonated high explosive
V_{corr}	Correcting function for the volume of the detonation products
V_{det}	Volume of detonation products
$V_{exp\,gas}$	Volume of gaseous products
v_W	Sound velocity in water or the acoustic velocity at the depth where the charge is positioned with allowance for the ambient water temperature
U	Internal energy
$U_{poducts}$	Internal energy of products
$U_{reactants}$	Internal energy of reactants
U_s	Internal energy of the isentropically expanded products
U_0	Internal energy of the isentropically unreacted explosive
$U(c)$	Internal energy of a specific compound in the condensed phase (solid or liquid)
$U^\theta(c)$	Internal energy of a specific compound in the condensed phase (solid or liquid) under standard conditions (298.15 K and 0.1 MPa pressure)
$U(g)$	Internal energy of a specific gaseous compound
$V_{cylinder\,wall}$	Cylinder wall velocity
W	Work
$W_{C\text{-}J}$	Velocity of gaseous products (fumes) at the C–J point
WPHE	Weight percent of high explosives
x_j	Mole fraction of the j-th component in the explosive mixture

y_i	Mole fraction of the i-th gaseous product
ZMWNI	Thermochemical computer code
$\Delta_f H^\theta$	Standard heat of formation of the desired species in the condensed phase (solid or liquid) or gas phase
$\Delta_f H^\theta(g)$	Standard heat of formation of a specific compound in the gas phase
$\Delta_f H^\theta(c)$	Standard heat of formation of a specific compound in the condensed phase (solid or liquid)
$\Delta_f H^\theta(\text{detonation product})_j$	Standard heat of formation of the jth detonation product
ΔH_c	Heat of combustion
ΔH_c^θ	Standard heat of combustion
ΔU_c	Energy of combustion
ΔU_c^θ	Standard energy of combustion
$\Delta V_{\text{Trauzl}}(\text{energetic compound})$	Volume of expansion for an explosive in the Trauzl lead block test
$\Delta V_{\text{Trauzl}}(\text{TNT})$	Volume of expansion for TNT in the Trauzl lead block test
α	Empirical constant of BKW-EOS
β	Empirical constant of BKW-EOS
κ	Empirical constant of BKW-EOS
δ	Dent depth
θ	Empirical constant of BKW-EOS
Ω	Oxygen balance
ρ_0	Initial (loading) density (g/cm^3)
ρ_{C-J}	Density at C–J point (g/cm^3)
ρ_W	Density of water at the gauge location site
γ	Adiabatic exponent
ω	Grüneisen coefficient or the second adiabatic coefficient
τ	Representative time of the process that is the time during which the pressure signal recorded drops from its maximum (at the front) to $P_m/e \approx 0.37 P_m$ where P_m is the peak pressure of the shock wave

About the authors

Mohammad Hossein Keshavarz, born in 1966, studied chemistry at Shiraz University and received his BSc in 1988. He also received a MSc and PhD at Shiraz University in 1991 and 1995. From 1997 until 2008, he was assistant professor, associate professor, and professor of physical chemistry at the University of Malek-ashtar in Shahin-shahr, Iran. Since 1997 he is lecturer and researcher at the Malek-ashtar University of Technology, Iran. He is the editor of two research journals in the Persian language. Keshavarz has published over 300 scientific papers in international peer reviewed journals, four book chapters, and eight books in the field of the assessment of energetic materials (four books in the Persian language).

Thomas M. Klapötke, born in 1961, studied chemistry at the TU Berlin and received his PhD in 1986 under the supervision of Hartmut Kopf. After his postdoctoral studies in Fredericton, New Brunswick with Jack Passmore he finished his habilitation at the TU Berlin in 1990. From 1995 until 1997 Klapötke was Ramsay professor of chemistry at the University of Glasgow in Scotland. Since 1997 he is professor and holds the chair of inorganic chemistry at LMU Munich. In 2009 Klapötke was appointed a visiting professor for mechanical engineering and chemistry at the Center of Energetic Concepts Development (CECD) at the University of Maryland (UM), College Park, and was also appointed a senior visiting scientist at the Energetics Technology Center (ETC) in La Plata, MD. In 2011 Klapötke was apointed an honorary fellow of the High Energy Materials Society of India (HEMSI). Klapötke is a fellow of the Royal Society of Chemistry (C. Sci., C. Chem. F.R.S.C., UK), a member of the American Chemical Society (ACS) and the Fluorine Division of the ACS, a member of the Gesellschaft Deutscher Chemiker (GDCh), a life member of the International Pyrotechnics Society (IPS), and a life member of the National Defense Industrial Association (NDIA, USA). Most of Klapötke's scientific collaborations are between LMU and the US Army Research Laboratory (ARL) in Aberdeen, MD and the US Army Armament Research, Development, and Engineering Center (ARDEC) in Picatinny, NJ. Klapötke also collaborates with the US Army Engineer Research and Development Center (ERDC) in Champaign, IL and several industrial partners in Germany and the USA. He is the executive editor of the Zeitschrift fur Anorganische and Allgemeine Chemie (ZAAC) and an editorial board member of Propellants, Explosives and Pyrotechnics (PEP), Journal of Energetic Materials and Central European Journal of Energetic Materials (CEJEM), and the International Journal of Energetic Materials and Chemical Propulsion (IJEMCP). Klapötke has published over 500 scientific papers in international peer-reviewed journals, 23 book chapters, and five books.

1 Heat of detonation

Combustion is a chemical reaction which is very fast, highly exothermic, is usually accompanied by a flame, and takes place between a substance and oxygen. The energy generated during combustion raises the temperature of the unreacted material and increases the rate of reaction. If the temperature of the combustible substance is raised above its ignition temperature, the heat which is released will be greater than the heat which is lost to the surrounding medium, and a flame will be observed. For organic energetic compounds such as propellants and explosives a large amount of gas at high temperatures is liberated during their combustion. In contrast to the combustion of fuels, the combustion of organic energetic compounds is a self-sustaining process which does not require the presence of oxygen in the surrounding atmosphere. An organic energetic compound can be classed as being a **deflagrating organic energetic compound** if a small amount of the compound in an unconfined space suddenly ignites on exposure to initiation sources such as a flame, spark, shock, friction, or high temperatures. Deflagrating organic energetic compounds burn faster and more violently than ordinary fuels, and with a flame, sparks, hissing, or crackling noise. When an organic energetic compound initiates and its decomposition occurs via a travelling shock wave rather than a thermal mechanism, it is called a **detonating organic explosive**. The detonation process for an organic energetic compound can be instigated either by burning to detonation or by an initial shock.

1.1 Basic knowledge of the heat of detonation

The **heat of reaction** is the net heat difference between the heat of formation of the reactants and products which participate in a chemical reaction. The **heat of combustion** is the heat of reaction for an oxidation reaction. An organic energetic compound can undergo either detonation as an explosive, or deflagration as a propellant. For deflagration (burning) of an energetic compound, the **heat of deflagration** can be used to represent the energy which is liberated. The **heat of detonation** shows the amount of energy that can be liberated on the detonation of explosives. However, the **heat of explosion**, denoted by Q_{expl}, is a general term that can be used to describe the quantity of heat which is released in the decomposition of an energetic compound which is acting as either an explosive or propellant. For both the detonation and deflagration processes, the heat liberated by these processes will raise the temperature of gaseous products. This is because the decomposition of an energetic compound is often extremely fast, which in turn causes the gaseous products to expand and release energy to the surroundings. However, the effectiveness of an energetic compound depends on the amount of energy available in it, and the rate of release of this available energy when detonation occurs. Q_{expl} is one of the most important thermodynamic parameters that determines the performance of explosives and propellants [1]. As shown in

https://doi.org/10.1515/9783110677652-001

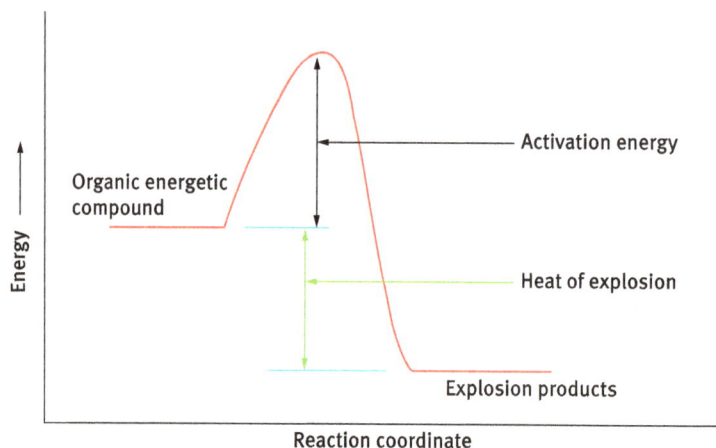

Figure 1.1: Energy profile of an explosive reaction.

Figure 1.1, Q_{expl} is the quantity of heat released when an explosive such as 1,3,5-tri-nitrotoluene undergoes detonation or deflagration. Since an organic energetic compound carries its own oxygen within the molecule, both detonation and deflagration occur even in the absence of external oxygen or air [2]. Q_{expl} is a quick and reliable criterion to evaluate the performance potential of a desired organic energetic compound [3]. For example, the higher the Q_{expl} of a rocket propellant, the higher its "specific impulse" will be, which is an index of propelling power of a rocket using a specific propellant [4].

1.1.1 Measurement of the heat of explosion

A bomb calorimeter is a commonly used instrument which can be used to estimate and compare the values of Q_{expl} for energetic materials held under constant volume (Figure 1.2). It involves burning the sample in a metal bomb containing an inert atmosphere (nitrogen-filled) while the bomb is kept submerged in a measured quantity of water within a thermal insulating chamber. The metal bomb has to withstand the large pressure within the calorimeter as the temperature of the water around the metal bomb is being measured. The measurement is completed after the products have cooled back to nearly room temperature by noting the increase in the temperature of the calorimeter water jacket. Q_{expl} is calculated from this temperature increase using the effective heat capacity of the calorimeter body and water jacket. The temperature rise of the bomb calorimeter must be recorded in order to calculate Q_{expl}. The values of Q_{expl} include contributions due to the cooling of the combustion products from the flame temperature to room temperature by considering phase changes such as water condensation.

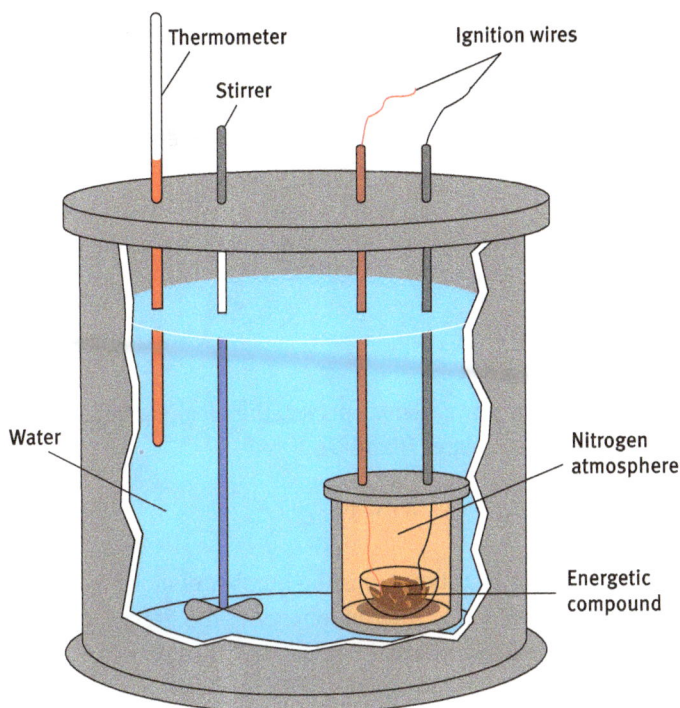

Figure 1.2: Bomb calorimeter for measuring the heat of explosion.

1.1.2 Heat of detonation and heat of formation

When an organic energetic compound is initiated to rapid burning and detonation, energy is released in the form of heat mainly due to oxidation reactions. Thus, Q_{expl} shows the heat which is released under adiabatic conditions and is a very important characteristic of an explosive, providing information about its work capacity. It can be expected that secondary high explosives and propellants possess high values of Q_{expl}. For propellants burning in gun chambers and secondary explosives in detonating devices, Q_{expl} is conventionally expressed in terms of constant volume conditions. For rocket propellants burning in the combustion chamber of a rocket motor under conditions of free expansion to the atmosphere, Q_{expl} is approximately used under constant pressure conditions. It should be noted that the reported values for the heat of explosion are a misnomer for three reasons:

(a) The term "heat" can be used to imply constant pressure enthalpy (H), whereas the internal energy (U) is an equally valid definition of "heat" in conditions of constant volume. Although a bomb calorimeter is a constant volume device, both of the terms H and U are used in the literature.

(b) A sample of an energetic compound usually undergoes combustion rather than detonation.

(c) For exothermic reactions, a negative sign should be used in accordance to the convention within the thermodynamics community. However, the energetic materials community reports Q_{expl} values as positive numbers.

The two thermodynamic quantities H and U are state functions for a given species, which can be related as follows:

$$H = U + PV, \tag{1.1}$$

where P is pressure and V is volume. For gaseous species, the ideal gas equation of state gives the following equation that relates H and U:

$$H(g) = U(g) + n(g)RT. \tag{1.2}$$

where R is the gas constant; T is temperature; $n(g)$ is the number of moles of gas involved; $H(g)$ and $U(g)$ are the enthalpy and internal energy of the desired gaseous species respectively. At room temperature, RT is equal to 2.48 kJ/mol. For condensed phase (liquid or solid) species, the value of the product of PV in equation (1.2) is small enough to allow the following approximation to be made:

$$H(c) \approx U(c), \tag{1.3}$$

where $H(c)$ and $U(c)$ are the enthalpy and internal energy of desired species in the condensed phase respectively. It can be assumed that under standard state conditions (298.15 K and 0.1 MPa pressure) $H^\theta = \Delta_f H^\theta$ for the species of interest relative to the reference elements at 298.15 K and 0.1 MPa pressure. Thus, equation (1.3) can be written as

$$U^\theta(c) \approx H^\theta(c) \approx \Delta_f H^\theta(c). \tag{1.4}$$

Since the enthalpies of formation are often more readily available than the internal energies of formation, in a good approximation, Q_{expl} can be calculated from the difference between the sum of the standard heats of formation of the products and the sum of the standard heats of formation of the reactants. In other words, Q_{expl} is simply the difference between $\Delta_f H^\theta$ of the products of explosion and $\Delta_f H^\theta$ of the explosive itself. The value of $\Delta_f H^\theta$ for chemical explosives may be calculated from knowledge of the individual bond energies between the atoms of an explosive molecule. Furthermore, the values of $\Delta_f H^\theta$ for different gaseous products are available in the literature. Thus, it is possible to calculate Q_{expl} from the value of $\Delta_f H^\theta$ of the assumed decomposition reaction. For secondary organic high explosives, Kamlet and Jacobs [5] used the term "heat of detonation", denoted by Q_{det}, for the calculation of the heat of detonation reaction as follows:

$$Q_{det} \approx \frac{-[\sum n_j \Delta_f H^\theta (\text{detonation product})_j - \Delta_f H^\theta (\text{explosive})]}{\text{formula weight of explosive}}, \tag{1.5}$$

where $\Delta_f H^\theta (\text{detonation product})_j$ and n_j are the standard heat of formation and the number of moles of the j-th detonation product, respectively. As seen in equation (1.5), a positive heat of formation (per unit weight) is favorable for an explosive, because this leads to a greater release of energy upon detonation and an improvement in performance. The assumed or computed equilibrium composition of the product gases can be used for evaluating the heat of formation of the detonation products. If the condensed heat of formation of the explosive and decomposition products of explosive are known, then using the standard heats of formation of the gaseous products will lead to the prediction of the heat of detonation of an explosive. Experimental data of $\Delta_f H^\theta$ for some pure and composite explosives are given in the Appendix.

1.1.3 Relationship between the heat of combustion and heat of formation

The energy of combustion is experimentally measured through the use of a bomb calorimeter filled with an excess of oxygen (Figure 1.3). For an energetic compound with general formula $C_a H_b N_c O_d$ in the solid or liquid phase, the combustion products would be liquid H_2O and gaseous CO_2 and N_2 on the basis of the following equation:

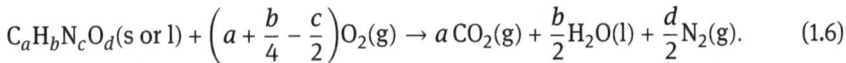

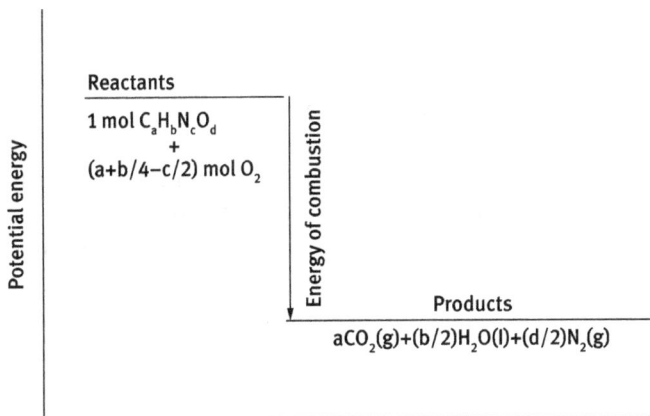

$$C_a H_b N_c O_d (\text{s or l}) + \left(a + \frac{b}{4} - \frac{c}{2}\right) O_2(g) \rightarrow a\, CO_2(g) + \frac{b}{2} H_2O(l) + \frac{d}{2} N_2(g). \tag{1.6}$$

Figure 1.3: The difference between the energy content of one mole of the desired energetic compound $C_a H_b N_c O_d$ and the energy of the combustion products formed on its oxidation by O_2.

The measured energy of combustion, ΔU_c, can be converted into the heat of combustion, ΔH_c, by the following equation:

$$\Delta H_c = \Delta U_c + \Delta n(g)RT = \Delta U_c + \left(\frac{d}{2} - \frac{b}{4} + \frac{c}{2}\right)RT, \tag{1.7}$$

where $\Delta n(g)$ is the change in the number of moles of gas for the reaction, and R is the gas constant. Thus, the standard heat of combustion in kJ/mol can be calculated by use of the equation

$$\Delta H_c^\theta = \Delta U_c^\theta + 2.4788\left(\frac{d}{2} - \frac{b}{4} + \frac{c}{2}\right). \tag{1.8}$$

Moreover, the standard heat of formation of the desired energetic compound C_aH_b N_cO_d can be calculated by the following equation [6]:

$$\Delta_f H^\theta(\text{energetic compound}) = a\Delta_f H^\theta[CO_2(g)] + \frac{b}{2}\Delta_f H^\theta[H_2O(l)] - \Delta H_c^\theta. \tag{1.9}$$

By assuming $\Delta_f H^\theta[CO_2(g)]$ and $\Delta_f H^\theta[H_2O(l)]$ to be equal to -393.5 and -285.8 kJ/mol, respectively, and using equations (1.8) and (1.9), the value of $\Delta_f H^\theta$ (energetic compound) can be obtained.

1.2 The assumed detonation products

Two different approaches are frequently used for predicting the detonation products. The first one is based on simple decomposition paths and the second approach is to use thermochemical computer codes.

1.2.1 Simple methods for the prediction of the detonation products

There are several different approaches such as the Kistiakowsky–Wilson rules, Modified Kistiakowsky–Wilson rules, and the Springall Roberts rules that are based on the distribution of oxygen atoms to carbon and hydrogen atoms through several steps to form the detonation products CO, CO_2, H_2O, and N_2 for $C_aH_bN_cO_d$ explosives [7]. Although these methods are simple in their approaches, it is difficult to derive decomposition pathways. Several simple methods have been developed to estimate the number of moles of detonation products formed in plausible decomposition pathways. Kamlet and Jacobs [5] have suggested that the overall stoichiometry of explosive decomposition products for an explosive having the general formula $C_aH_bN_cO_d$ may be given as

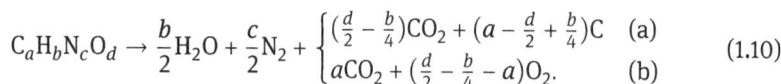

$$C_aH_bN_cO_d \rightarrow \frac{b}{2}H_2O + \frac{c}{2}N_2 + \begin{cases} (\frac{d}{2} - \frac{b}{4})CO_2 + (a - \frac{d}{2} + \frac{b}{4})C & \text{(a)} \\ aCO_2 + (\frac{d}{2} - \frac{b}{4} - a)O_2. & \text{(b)} \end{cases} \tag{1.10}$$

Kamlet and Jacobs [5] assumed that $C_aH_bN_cO_d$ high explosives generally have crystal densities ranging from 1.7 to 1.9 g/cm^3, and the explosives which are used have densities close to their theoretical maximum density. They argue that for explosives at these densities, the product compositions given in equation (1.6) can be called the "H_2O–CO_2 arbitrary".

For organic $C_aH_bN_cO_dF_eCl_f$ explosives, it can be assumed that all nitrogen contained in the explosive forms N_2, fluorine is converted to HF, chlorine to HCl, a portion of the oxygen forms H_2O, and carbon is preferentially oxidized to CO rather than CO_2. The following pathways can then be written to obtain the detonation products:

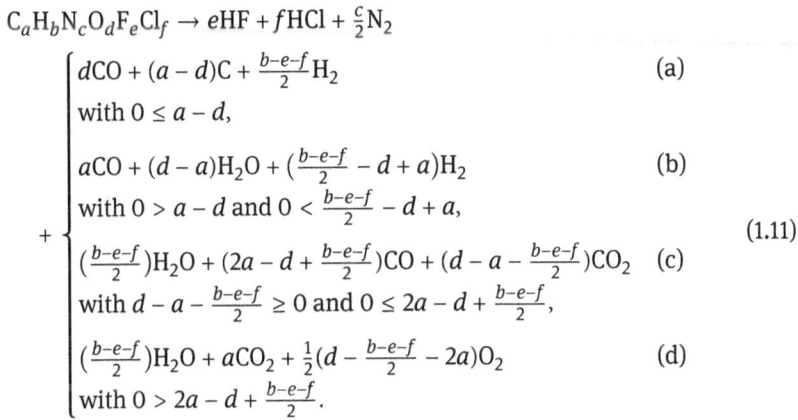

$$C_aH_bN_cO_dF_eCl_f \rightarrow eHF + fHCl + \tfrac{c}{2}N_2$$

$$+ \begin{cases} dCO + (a-d)C + \frac{b-e-f}{2}H_2 & \text{(a)} \\ \text{with } 0 \le a-d, \\[4pt] aCO + (d-a)H_2O + (\frac{b-e-f}{2} - d + a)H_2 & \text{(b)} \\ \text{with } 0 > a-d \text{ and } 0 < \frac{b-e-f}{2} - d + a, \\[4pt] (\frac{b-e-f}{2})H_2O + (2a - d + \frac{b-e-f}{2})CO + (d - a - \frac{b-e-f}{2})CO_2 & \text{(c)} \\ \text{with } d - a - \frac{b-e-f}{2} \ge 0 \text{ and } 0 \le 2a - d + \frac{b-e-f}{2}, \\[4pt] (\frac{b-e-f}{2})H_2O + aCO_2 + \frac{1}{2}(d - \frac{b-e-f}{2} - 2a)O_2 & \text{(d)} \\ \text{with } 0 > 2a - d + \frac{b-e-f}{2}. \end{cases}$$

(1.11)

For some $C_aH_bN_cO_d$ explosives, the decomposition paths shown in equation (1.11) give a more reliable prediction of Q_{det} compared to equation (1.10). Two different values can be obtained for Q_{det}, because two different phases (gas and liquid) can be assumed for water, which is one of the major detonation products. To demonstrate the use of equations (1.10) and (1.11) for the calculation of Q_{det}, cyclotetramethylenetetranitramine (HMX) with the empirical formula $C_4H_8N_8O_8$ is chosen. Since equation (1.10) (b) can only be used for organic explosives containing enough oxygen to convert all hydrogen and carbon atoms to H_2O and CO_2, respectively, there are only a few explosives such as nitroglycerine (NG: $C_3H_5N_3O_9$) that can follow this decomposition path. Therefore, equation (1.10) (a) is used for most organic explosives. Thus, equation (1.10) (a) gives the following detonation products and heat of detonation:

$$C_4H_8N_8O_8 \rightarrow 4N_2 + 4H_2O + 2CO_2 + C$$

$$Q_{det}[H_2O(g)] \cong \frac{-[4\Delta_f H^\theta(H_2O(g)) + 2\Delta_f H^\theta(CO_2) - \Delta_f H^\theta(HMX)]}{\text{formula weight of HMX}}$$

$$= \frac{-[(4)(-241.8\,kJ/mol) + (2)(-393.5\,kJ/mol) - (74.8\,kJ/mol)]}{296\,g/mol}$$

$$= 6.18\,kJ/g,$$

$$Q_{det}[H_2O(l)] \approx \frac{-[4\Delta_f H^\theta(H_2O(l)) + 2\Delta_f H^\theta(CO_2) - \Delta_f H^\theta(HMX)]}{\text{formula weight of HMX}}$$

$$= \frac{-[(4)(-285.8\,kJ/mol) + (2)(-393.5\,kJ/mol) - (74.8\,kJ/mol)]}{296\,g/mol}$$

$$= 6.77\,kJ/g.$$

Meanwhile, the decomposition path shown in equation (1.11)(c) should be used because the condition

$$d \geq a + \frac{b-e-f}{2} \quad \left(8 \geq 4 + \frac{8-0-0}{2}\right)$$

is satisfied for HMX:

$$C_4H_8N_8O_8 \rightarrow 4N_2 + 4H_2O + 4CO.$$

However, the heat of detonation can be calculated on the basis of the new detonation products as follows:

$$Q_{det}[H_2O(g)] \approx \frac{-[4\Delta_f H^\theta(H_2O(g)) + 4\Delta_f H^\theta(CO) - \Delta_f H^\theta(HMX)]}{\text{formula weight of HMX}}$$

$$= \frac{-[(4)(-241.8\,kJ/mol) + (4)(-110.5\,kJ/mol) - (74.8\,kJ/mol)]}{296\,g/mol}$$

$$= 5.02\,kJ/g,$$

$$Q_{det}[H_2O(l)] \approx \frac{-[4\Delta_f H^\theta(H_2O(l)) + 4\Delta_f H^\theta(CO) - \Delta_f H^\theta(HMX)]}{\text{formula weight of HMX}}$$

$$= \frac{-[(4)(-285.8\,kJ/mol) + (2)(-110.5\,kJ/mol) - (74.8\,kJ/mol)]}{296\,g/mol}$$

$$= 5.61\,kJ/g.$$

The necessary values from experimental data for H_2O, CO, and CO_2 were taken from the NIST Chemistry WebBook [8]. To compare the effect of the different detonation products formed in equations (1.10) and (1.11), the heats of detonation for some well-known $C_aH_bN_cO_d$ organic explosives which were calculated using these equations are given in Table 1.1. The root mean square (RMS) of deviations are also given in Table 1.2, and which can be defined as

$$\text{RMS deviation (kJ/g)} = \sqrt{\frac{1}{N}\sum_{i=1}^{N}\text{Dev}_i^2}, \qquad (1.12)$$

where N represents the number of heat of detonation measurements which were included.

Table 1.1: Comparison of the predicted and experimental heats of detonation of some well-known $C_aH_bN_cO_d$ explosives.

Explosive[a]	Δ_fH^θ (c)[b] (kJ/mol)	$Q_{det}[H_2O(g)]$ (kJ/g)			$Q_{det}[H_2O(l)]$ (kJ/g)		
		Exp.[c]	Eq. (1.11)[d]	Eq. (1.10)[d]	Exp.[c]	Eq. (1.11)[d]	Eq. (1.10)[d]
HMX ($C_4H_8N_8O_8$)	75.0	5.732 [9]	5.017 (0.715)	6.180 (−0.448)	6.192 [1]	5.611 (0.581)	6.774 (−0.582)
RDX ($C_3H_6N_6O_6$)	66.9	5.941 [9]	5.038 (0.903)	6.201 (−0.26)	6.318 [1]	5.636 (0.682)	6.799 (−0.481)
TNT ($C_7H_5N_3O_6$)	67.1	4.268 [9]	2.644 (1.624)	5.418 (−1.15)	4.561 [1]	2.644 (1.917)	5.904 (−1.343)
PETN ($C_5H_8N_4O_{12}$)	−538.9	5.732 [9]	5.791 (−0.059)	6.335 (−0.603)	6.318 [1]	6.351 (−0.033)	6.891 (−0.573)
TETRYL ($C_7H_5N_5O_8$)	20.0	4.56 [9]	3.607 (0.953)	5.941 (−1.381)	4.770 [1]	3.761 (1.009)	6.326 (−1.556)
DATB ($C_6H_5N_5O_6$)	−98.7	3.81 [9]	2.322 (1.488)	4.912 (−1.102)	4.10 [10]	2.322 (1.778)	5.368 (−1.268)
NQ ($CH_4N_4O_2$)	−92.9	2.732 [1]	2.498 (0.234)	3.761 (−1.029)	3.071 [1]	2.920 (0.151)	4.607 (−1.536)
TATB ($C_6H_6N_6O_6$)	−139.7	—	1.975	4.502	3.063 [1]	1.975 (1.088)	5.012 (−1.949)
NM (CH_3NO_2)	−113.0	4.301 [9]	3.925 (0.377)	5.703 (−1.402)	4.820 [1]	4.644 (0.176)	6.786 (−1.966)
RMS deviation (kJ/g)			0.954	1.006		1.048	1.365

a See the Appendix for a glossary of compound names and chemical formulas.
b Heat of formation of pure explosives were obtained from [1].
c References for Q_{det} are given in brackets.
d Differences of the measured and calculated data on the basis of the decomposition paths shown in equations (1.10) and (1.11) are given in parentheses.

1.2.2 Prediction of the detonation products on the basis of computer codes and using quantum mechanical calculations for the prediction of Q_{det}

Different thermochemical computer codes such as CHEETAH [11] and EXPLO5 [10] can be used to predict the value of Q_{expl} for an energetic compound. For example, EXPLO5 [10] is a computer program for the calculation of the detonation parameters of explosives. It is based on the chemical equilibrium and steady-state model of detonation. It uses the Becker–Kistiakowsky–Wilson (BKW) equation of state, which is based upon a repulsive potential applied to the virial equation of state for expressing the state of gaseous detonation products [12]. The BKW equation has the following form [12]:

$$\frac{PV}{RT} = 1 + \left(\frac{\kappa \sum y_i k_i}{[V(T + \theta)]^\alpha} \right) \exp\left[\beta\left(\frac{\kappa \sum y_i k_i}{[V(T + \theta)]^\alpha} \right) \right], \tag{1.13}$$

where P is the pressure, V is the volume, R is the gas constant, T is the temperature, y_i is the mole fraction of the i-th gaseous product, k_i are the molar covolumes of the i-th gaseous product, and α, β, κ, and θ are empirical constants. EXPLO5 minimizes the free energy through a suitable technique, which was developed by White, Johnson, and Dantzig [13] and modified by Mader [12]. This technique is applied to the mathematical expression of the equilibrium state of the detonation products. According to this technique, the system of equations is formed and solved by applying modified Newton's method, but with some approximations [10].

For CHEETAH [11] calculations, it should be mentioned that the heats of formation are included in the library of reactants in this suite of programs and consist of values compiled from the literature or passed along by oral tradition. The user's manual gives the estimated errors in the heats of formation. The Chapman–Jouguet (C–J) state is calculated for the designated explosive, while the heats of detonation predicted using CHEETAH [11] and the Jacobs–Cowperthwaite–Zwisler (JCZS) equation of state (EOS), or JCZS-EOS library (S for Sandia) [14] are obtained by executing the "Standard Detonation Run", in which the adiabatic expansion of the product gases from the C–J state to 1 atm is calculated. In these calculations, the value of Q_{det} corresponds to the energy difference between the reactants and all of the products at the end of this expansion. Rice and Hare [15] have predicted the product concentrations using the thermochemical code CHEETAH [10] and the JCZS product library [14] for 34 $C_aH_bN_cO_d$ explosives. The results indicated that 94 % of the gaseous products consist of only five compounds: H_2O, H_2, N_2, CO, and CO_2. More than 97 % of the gaseous products consist of only the above mentioned five species for 30 out of the 34 explosives which were investigated. Rice and Hare [15] assumed that the detonation products are formed according to the following decomposition equation, since CO is predicted to be a major component of the product gases by the thermochemical calculations:

$$C_aH_bN_cO_d \rightarrow eH_2O + fN_2 + gCO_2 + hCO + iH_2 + jC + k(\text{other products}), \tag{1.14}$$

where the number of moles of products e, f, g, h, i, j, and k are given by the CHEETAH/ JCZS calculations. They assumed that the contribution of k in equation (1.14) is small and referred to this as the modified Kamlet and Jacobs method [15]. They predicted the heat of formation for explosives in the condensed phase using quantum mechanical calculations and the general interaction properties function (GIPF) methodology. They also calculated the heats of detonation using quantum mechanically predicted values for the condensed phase heats of formations of the explosives, and the experimental values of the heats of formation for CO_2, CO, and H_2O. The standard heats of formation of C, H_2, and N_2 in equation (1.14) are zero. For pure explosives, they have indicated that the quantum mechanical based method using the modified Kamlet and Jacobs method is in better agreement with experimental values than the H_2O–CO_2 arbitrary [15]. For $C_aH_bN_cO_d$ explosives, equation (1.14) and the H_2O–CO_2 arbitrary were used to calculate the heat of detonation of pure explosives and explosive formulations [15].

1.3 New empirical methods for the prediction of Q_{det} without considering the detonation products

In recent years, different correlations have been developed to predict the values of Q_{det} for secondary $C_aH_bN_cO_d$ high explosives, without considering their detonation products. These methods are reviewed here.

1.3.1 Using the gas and condensed phase heats of formation of explosives

1.3.1.1 Elemental composition and standard gas phase heat of formation of an explosive

For aromatic and nonaromatic, pure secondary high explosives of the general formula $C_aH_bN_cO_d$, it was shown that the following equation can be used to estimate $Q_{det}[H_2O(l)]$ [16]:

$$Q_{det}[H_2O(l)]_{aromatic}$$
$$= \frac{61.78a - 51.32b + 30.66c + 91.45d - 0.0667[\Delta_f H^\theta(g) \text{ of explosive}]}{\text{formula weight of explosive}}, \quad (1.15)$$

$$Q_{det}[H_2O(l)]_{nonaromatic}$$
$$= \frac{58.72a - 55.01b - 21.23c + 250.9d + 1.065[\Delta_f H^\theta(g) \text{ of explosive}]}{\text{formula weight of explosive}}, \quad (1.16)$$

where $\Delta_f H^\theta(g)$ is the standard gas phase heat of formation of the explosive in kJ/mol, which can be estimated by different methods, e. g. quantum mechanical, group additivity, and empirical methods [17]. It should be noted that in these equations no prior knowledge of any measured, estimated, or calculated physical, chemical, or thermochemical properties of the explosive and its assumed detonation products is required other than the easily calculated gas-phase heat of formation. This new procedure, which is based on the calculated gas-phase heats of formation, shows surprisingly very good agreement with experimental values. Since the values of the coefficients of $\Delta_f H^\theta(g)$ are small relative to the coefficients of elemental composition in equations (1.15) and (1.16), the contributions of the last term in these equations are low. Thus, there is no need to use more reliable complex quantum mechanical methods for the calculation of $\Delta_f H^\theta(g)$ of explosives here.

Two examples are given here to illustrate the application of the method for aromatic and nonaromatic high explosives. Hexanitrohexaazaisowurtzitane, also called HNIW or CL-20, is a nitramine explosive with the formula $C_6H_6N_{12}O_{12}$, and is the most powerful explosive being investigated at the pilot scale or larger [18]. Hexanitrostilbene (HNS) is an organic compound with the formula $C_{14}H_6N_6O_{12}$, which is used as a heat-resistant high explosive. If Benson's group additivity method [19] is used – which can be easily applied through knowledge of the molecular structure of a desired explosive and using the NIST Chemistry WebBook [8] – the predicted $\Delta_f H^\theta(g)$ value of 150 kJ/mol is obtained for HNS. For CL 20, the group additivity method cannot be used because of the absence of a specific group. However, the value of $\Delta_f H^\theta(g)$ for CL 20 was predicted to be about 500 kJ/mol on the basis of isodesmic reaction calculations using the G4(MP2) and B3LYP/cc-pVTZ levels of theory [20]. Thus, equations (1.15) and (1.16) give the following values for $Q_{det}[H_2O(l)]$ of HNS and CL 20 as aromatic and nonaromatic explosives, respectively:

$$Q_{det}[H_2O(l)]_{aromatic} = \frac{61.78(14) - 51.32(6) + 30.66(6) + 91.45(12) - 0.0667(150)}{450.23}$$
$$= 4.06 \text{ kJ/g,}$$

$$Q_{det}[H_2O(l)]_{nonaromatic} = \frac{58.72(6) - 55.01(6) - 21.23(12) + 250.9(12) + 1.065(500)}{438.18}$$
$$= 7.56 \text{ kJ/g.}$$

The experimentally obtained values for CL 20 and HNS are 6.234 [21] and 4.088 kJ/g [1], respectively, which are consistent with the predicted values.

1.3.1.2 The corrected Kamlet and Jacobs method for aromatic explosives

For aromatic energetic compounds, it was found that the results predicted by the Kamlet and Jacobs method [5] are overestimated in comparison with the experimental data [22]. However, it was shown that the following equation provides more reliable predictions in comparison to the Kamlet and Jacobs method [5, 21]:

$$Q_{det}[H_2O(l)]_{aromatic} = -2.181 + \frac{45.6b + 201.2d + \Delta_f H^\theta(\text{explosive})}{\text{formula weight of explosive}}. \quad (1.17)$$

As can be seen in Table 1.2, the results of this equation are in good agreement with the measured data.

1.3.1.3 The correctedKamlet and Jacobs method for aromatic and nonaromatic explosives

It has been found that an appropriate correlation of the following form can provide a suitable pathway to obtain reliable predictions of the heats of detonation for non-aromatic and aromatic explosives [22]:

$$Q_{det}[H_2O(l)]_{nonaromatic} = 2.111 + 0.915Q_{H_2O-CO_2} - 4.584(a/d) - 0.464(b/d), \quad (1.18)$$

$$Q_{det}[H_2O(l)]_{aromatic} = -1.965 + 0.993Q_{H_2O-CO_2} + 0.029(a/d) - 0.106(b/d), \quad (1.19)$$

where $Q_{H_2O-CO_2}$ is the heat of detonation based on the "H_2O-CO_2 arbitrary" where water is in liquid state. Equations (1.18) and (1.19) provide a reliable correlation for a quick estimation of the heats of detonation for a wide range of energetic materials including under-oxidized and over-oxidized, nonaromatic explosives. Table 1.2 compares the predicted results of these correlations with experimental data.

1.3.2 Using structural parameters of high explosives

1.3.2.1 Nonaromatic explosives

For nonaromatic explosives, it was found that the following equation can provide a suitable pathway to predict the value of the heat of detonation [23]:

$$Q_{det}[H_2O(l)]_{nonaromatic} = 5.081 + 0.836(d/a) - 1.604(b/d) + 2.727C_{SSP}, \quad (1.20)$$

where d/a and b/d are the ratios of oxygen to carbon and hydrogen to oxygen atoms, respectively; C_{SSP} is the contribution of some specific structural parameters in a nonaromatic $C_aH_bN_cO_d$ explosive. The values of C_{SSP} can be given as follows:
(a) for cyclic nitramine, $C_{SSP} = 0.35$;
(b) for a nonaromatic explosive that has the $-N-C(=O)-N-$ functional group, the value of C_{SSP} is -1.0 if it does not contain more than two nitro groups.

As is indicated in Table 1.2, the predicted values of $Q_{det}[H_2O(l)]$ for several nonaromatic, energetic compounds are calculated and compared with the experimental values.

Table 1.2: Comparison of the predicted and experimental heats of detonation of some aromatic and nonaromatic $C_a H_b N_c O_d$ explosives.

Explosive	$\Delta_f H^\theta$ (c) [1] (kJ/mol)	Exp. [1]	Eq. (1.17)	Eqs. (1.18) and (1.19)	Eq. (1.20)	Eq. (1.21)	Eq. (1.22)
2,4,6-Trinitrophenol ($C_6H_3N_3O_7$)	−248.4	3.437	3.456	3.487	—	3.751	3.231
2,4,6-Trinitroaniline ($C_6H_4N_4O_6$)	−84.0	3.589	3.537	3.544	—	3.822	4.004
1-Methoxy-2,4,6-trinitrobenzene ($C_7H_5N_3O_7$)	−153.2	3.777	3.908	3.899	—	3.651	3.975
2,4,6-Trinitrobenzoic acid ($C_7H_3N_3O_8$)	−403.0	3.008	3.009	3.059	—	2.757	3.235
3-Methyl-2,4,6-trinitrophenol ($C_7H_5N_3O_7$)	−252.3	3.370	3.491	3.494	—	3.651	3.975
Ethyl nitrate ($C_2H_5NO_3$)	−190.2	4.154	—	4.240	3.66	—	3.991
Nitroethane ($C_2H_5NO_2$)	−134.0	1.686	—	1.797	1.91	—	2.859
Nitrourea ($CH_3N_3O_3$)	−282.6	3.745	—	3.785	3.26	—	4.482
Urea nitrate ($CH_5N_3O_4$)	−134.4	3.211	—	3.484	3.69	—	3.789
1,2,3-Propanetriol trinitrate ($C_3H_5N_3O_9$)	−370.7	6.671	—	6.764	6.70	—	6.520

1.3.2.2 Aromatic explosives

For various aromatic explosives, the general form of the correlation based on the structural parameters can be given as follows [24]:

$$Q_{det}[H_2O(l)]_{aromatic} = 2.129 + 0.178c + 0.874(d/a) + 0.160(b/d)$$
$$+ 0.965C_{SFG}, \tag{1.21}$$

where C_{SFG} is the contribution of some specific functional groups in aromatic C_aH_b N_cO_d energetic compounds. The value of C_{SFG} is –1.0 for aromatic energetic compounds that have some specific functional groups, namely: –COOH, NH_4^+, two –OH (or one –OH with one –NH$_2$) and three –NH$_2$. Comparison of the predicted results obtained using this method with the experimental data is given in Table 1.2.

1.3.2.3 General correlation for both aromatic and nonaromatic explosives

It was indicated that the ratios of oxygen to carbon and hydrogen to oxygen atoms, the presence of cyclic nitramines and the contributions of some specific polar functional groups can be used to predict $Q_{det}[H_2O(l)]$ as follows [25]:

$$Q_{det}[H_2O(l)] = 3.198 + 1.223(d/a) - 0.625(b/d)$$
$$+ 1.193P_{cyc,nitramine} - 1.408C_{polar}, \tag{1.22}$$

where $P_{cyc,nitramine}$ is a correcting function that can be applied only for cyclic nitramines; C_{polar} is the contribution of some specific polar or functional groups in aromatic and nonaromatic $C_aH_bN_cO_d$ energetic compounds. For cyclic nitramines such as 1,3,5-trinitro-1,3,5-triazacyclohextane (RDX), the value of $P_{cyc,nitramine}$ is 1.0. Different values are expected for C_{polar}, which can be specified as follows:

(1) Nitroaromatics: The value of C_{polar} is equal to 0.8 for aromatic energetic compounds that have some specific polar functional groups, namely one –COOH, –O$^-$, two –OH (or one –OH with one –NH$_2$) or three –NH$_2$ groups.

(2) Nonaromatic energetic compounds: For nitramines containing the polar functional group –NH–NO$_2$, C_{polar} equals 1.25. The value C_{polar} is 2.5 if the nitrate salt of the amino group (–NH$_2$·HNO$_3$) is present.

If the two above conditions for aromatic and nonaromatic energetic compounds are not satisfied, the value of C_{polar} equals zero.

1.3.3 Prediction of the heat of explosion in double-base and composite modified double-base propellants

Double-base propellants are known as smokeless propellants, which contain nitrocellulose (NC) and nitroglycerine (NG) as two major components of these propellants.

Some additives such as burning rate catalysts, modifiers, and anti-aging agents are added to the compositions of double-base propellants. These additives may result in superior mechanical properties at high and low environmental temperatures, as well as improving the burning rate characteristics. Since NG is highly shock sensitive, other types of nitrate esters such as ethyleneglycol dinitrate (EGDN), theriethyleneglycol dinitrate (TEGDN), and trimethylolethane trinitrate (TMETN) can be used to formulate non-NG double-base propellants [4].

For composite modified double base (CMDB) propellants, crystalline ammonium perchlorate (AP), cyclotetramethylene tetranitramine (HMX), cyclotrimethylene trinitramine (RDX), or Al particles are mixed with nitro polymers in order to increase the energy of double-base propellants. However, the physiochemical and performance characteristics of CMDB propellants are intermediate between those of composite and double-base propellants. Since CMDB propellants have great potential in producing a high specific impulse and flexibility of the burning rate [4, 7], they are widely used. For double-base and CMDB propellants, the knowledge of the Q_{expl} of propellants is especially important, because the heat of explosion is directly proportional to the explosion temperature and indirectly related to other performance characteristics such as powder force, powder potential, and the specific impulse. For the calculation of the Q_{expl} of double-base and CMDB propellants, two models were developed, namely: artificial neural network (ANN) and multiple linear regression (MLR) models [26]. For double-base and CMDB propellants, it was shown that the predicted results on the basis of the ANN and MLR models are more reliable than those obtained by the mass percent and heat of explosion of individual components [26]. For double-base propellants, the %NC and %NG values are the most important parameters affecting the Q_{expl} of double-base propellants. For CMDB propellants, the percentage of energetic plasticizer and percentage of nitramine are the most important parameters. Although these methods are complex, they have some advantages:

(i) there is no need to use the heats of formation of different species;
(ii) they do not require knowledge of the combustion products;
(iii) it is not restricted to $C_aH_bN_cO_d$ propellants, and can be applied to double-base and CMDB propellants containing other elements;
(iv) they are more simple to use in comparison with the other computer codes.

1.4 Calculation of the heat of detonation (explosion) temperature of ideal and non-ideal energetic compounds

There is a continuing need for the ability to be able to reliably predict the detonation velocities and other detonation parameters. The C–J theory has traditionally been used for this purpose since thermodynamic equilibrium of detonation products is reached instantaneously. Thus, an **ideal explosive** is one in which its performance can be described adequately for engineering purposes by steady-state detonation calculations using appropriate EOSs. Ideal explosives should show a short reaction zone and have small failure diameters which are suitable for practical applications. In contrast to ideal explosives, **non-ideal explosives** are often poorly modeled by the C–J theory, because they have slow chemical reaction rates with respect to the hydrodynamic time scale. The C–J assumption of instantaneous thermodynamic equilibrium breaks down, so that the detonation velocity of non-ideal explosives varies sharply from the C–J values.

Since the combustion of aluminum particles in explosives occurs behind the reaction front during the expansion of the gaseous detonation products, aluminum particles do not participate in the reaction zone, but act as inert ingredients. It is not clear what degree of aluminum is oxidized at the C–J point for a mixture of high explosives with aluminum. Thermodynamic calculations of detonation parameters in thermochemical computer codes are carried out by assuming a certain degree of oxidation of aluminum. Powdered aluminum appears to be completely reacted near the C–J point, whereas powdered AN does not. Both aluminum and AN increase the heat of detonation as they react, but aluminum raises the temperature of the products, whereas AN lowers it. Since the product molecules from the burning of aluminum are Al_2O_3 – but for AN are H_2O and N_2 – the particle density of the detonation products of AN increases, whereas aluminum produces products that lower it. Increasing particle density can shift the energy from the thermal to intermolecular potential. Since the decomposition of AN lowers the temperature, this situation determines how much AN is decomposed near the C–J point. Meanwhile, the burning of aluminum raises the temperature, which increases the burn-rate of aluminum.

For explosives containing aluminum (Al) and/or ammonium nitrate (AN), the behavior of Al and/or AN explosives cannot be described by the steady-state detonation calculations. Since there is the physical separation of fuel and oxidizer in these explosives, secondary reactions occur between explosion products, and the chemical reaction zone spreads. For calculation of the non-ideal behaviors of these explosives, computer codes may be used by assuming the partial equilibrium [27]. Some semiempirical models have also been developed, based on the partial consumption of Al and AN for estimation of the detonation velocity and pressure of ideal and non-ideal explosives containing Al and AN [27–36].

The use of finely dispersible powders of Al in composite explosives can increase their performance because they can enhance the heat of explosion. Aluminized explosives enhance not only air blast but also the bubble energies in underwater weapons. Moreover, they raise the reaction temperature and incendiary effects. The combustion of Al particles in explosives occurs behind the reaction front during the expansion of the gaseous detonation products. Thus, Al particles do not participate in the reaction zone but act as inert ingredients. For thermodynamic calculations of detonation parameters by thermochemical computer codes, a certain degree of oxidation of Al should be assumed because it is not clear what degree of Al is oxidized at the C–J point for aluminized explosives. The presence of Al and AN in composite explosives shows different behaviours [37]:

(1) In contrast to powdered AN, powdered Al may react completely near the C–J point.
(2) The heat of detonation can be increased by the reaction of both Al and AN. The reaction of Al raises the temperature of the products, but AN lowers it.
(3) The reaction of Al produces Al_2O_3, but AN provides H_2O and N_2. Thus, the particle densities of the detonation products of AN and Al may be increased and decreased, respectively. It can be expected that increasing particle density in AN can shift the energy of detonation products from the thermal to intermolecular potential.
(4) Since the decomposition of AN lowers the temperature, the amount of decomposition of AN can be determined near the C–J point. The burning of Al raises the temperature, which can increase the burn-rate of Al.

It is essential to find the equilibrium composition of the detonation products for prediction of detonation or combustion properties of energetic compounds. The free energy minimization technique is used to estimate the equilibrium composition of the detonation products in thermochemical computer codes [2]. It was recently shown that the decomposition paths given in equation (1.11) can be extended to also consider the detonation products of ideal $C_aH_bN_cO_dF_eCl_f$ explosives as well as nonideal aluminized and AN explosives in order to allow calculation of the detonation velocity [56]. Equation (1.23) shows the appropriate decomposition paths for various explosives with the general formula $C_aH_bN_cO_dF_eCl_fAl_g(NH_4NO_3)_h$, in which the percent participation of aluminum and AN in these reactions depends on the oxygen content of the other ingredients. Some of the aluminum will react with oxygen-rich detonation products such as H_2O to form Al_2O_3. Some of the AN decomposes to produce N_2, H_2O, and O_2, and the oxygen molecules which are produced can react with oxygen-deficient detonation products. As in equation (1.11), it was also assumed that all nitrogen is converted into N_2, fluorine to HF, chlorine to HCl, while a portion of the oxygen atoms form H_2O, with carbon atoms being preferentially oxidized to CO rather than CO_2. A list of ideal $C_aH_bN_cO_dF_eCl_f$ explosives as well as nonideal aluminized and AN explosives versus their compositions and the condensed phase heat of formation are given in the Appendix:

$$\xrightarrow[\substack{(78\,\% \text{ AN reacted})}]{a=b=c=d=e=f=g=0,\ h=1\ (\text{pure AN})} 0.78\,N_2(g) + 1.56\,H_2O + 0.39\,O_2(g)$$
$$+ 0.22\,NH_4NO_3(s) \tag{a}$$

$$\xrightarrow[\substack{(97\,\% \text{ Al, }93\,\% \text{ AN reacted})}]{a=b=c=d=e=f=0\ (\text{AN + Al})} 0.93h\,N_2(g) + 1.455g\,H_2(g)$$
$$+ (1.86h - 1.455g)H_2O + 0.465h\,O_2(g)$$
$$+ 0.485g\,Al_2O_3(s) + 0.03g\,Al(s)$$
$$+ 0.07h\,NH_4NO_3(s) \tag{b}$$

$$\xrightarrow[\substack{(25\,\% \text{ Al, }10\,\% \text{ AN reacted})}]{d \le a+0.375g-0.3h} e\,HF(g) + f\,HCl(g) + (\tfrac{c}{2} + 0.1h)N_2(g)$$
$$+ (d + 0.3h - 0.375g)CO(g)$$
$$+ (a - d - 0.3h + 0.375g)C(s)$$
$$+ (\tfrac{b-e-f}{2} + 0.2h)H_2(g) + 0.125g\,Al_2O_3(s)$$
$$+ 0.75g\,Al(s) + 0.9h\,NH_4NO_3(s) \tag{c}$$

$$\xrightarrow[\substack{(36\,\% \text{ Al, }13\,\% \text{ AN reacted})}]{a+0.54g-0.39h<d<a+\frac{b-e-f}{2}+0.54g-0.78h} e\,HF(g) + f\,HCl(g) + (\tfrac{c}{2} + 0.13h)N_2(g)$$
$$+ a\,CO(g) + (d + 0.39h - a - 0.54g)H_2O$$
$$+ (\tfrac{b-e-f}{2} - 0.13h - d + a + 0.54g)H_2(g) \tag{1.23}$$
$$+ 0.18g\,Al_2O_3(s) + 0.64g\,Al(s)$$
$$+ 0.87h\,NH_4NO_3(s) \tag{d}$$

$$\xrightarrow[\substack{(30\,\% \text{ Al, }15\,\% \text{ AN reacted})}]{a+\frac{b-e-f}{2}+0.45g-0.9h \le d<2a+\frac{b-e-f}{2}-0.9h+0.45g} e\,HF(g) + f\,HCl(g) + (\tfrac{c}{2} + 0.15h)N_2(g)$$
$$+ (\tfrac{b-e-f}{2} - 0.45g)H_2O(g)$$
$$+ (2a - d + \tfrac{b-e-f}{2} - 0.45h)CO(g)$$
$$+ (d - a - \tfrac{b-e-f}{2} + 0.45h)CO_2(g)$$
$$+ (0.45g + 0.3h)H_2(g)$$
$$+ 0.15g\,Al_2O_3(s) + 0.7g\,Al(s)$$
$$+ 0.85h\,NH_4NO_3(s) \tag{e}$$

$$\xrightarrow[\substack{(30\,\% \text{ Al, }15\,\% \text{ AN reacted})}]{d \ge 2a+\frac{b-e-f}{2}-0.9h+0.45g} e\,HF(g) + f\,HCl(g) + (\tfrac{c}{2} + 0.15h)N_2(g)$$
$$+ (\tfrac{b-e-f}{2} + 0.3h)H_2O(g) + a\,CO_2(g)$$
$$+ (\tfrac{d}{2} + 0.075h - a - \tfrac{b-e-f}{4} - 0.225g)O_2(g)$$
$$+ 0.15g\,Al_2O_3(s) + 0.7g\,Al(s)$$
$$+ 0.85h\,NH_4NO_3(s). \tag{f}$$

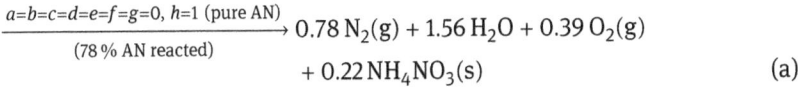

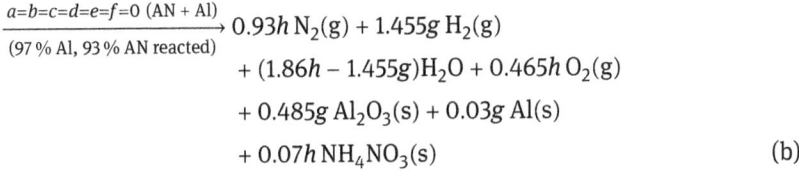

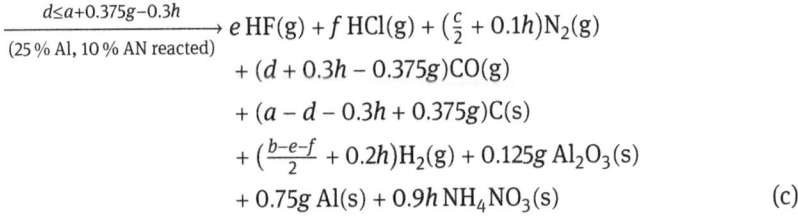

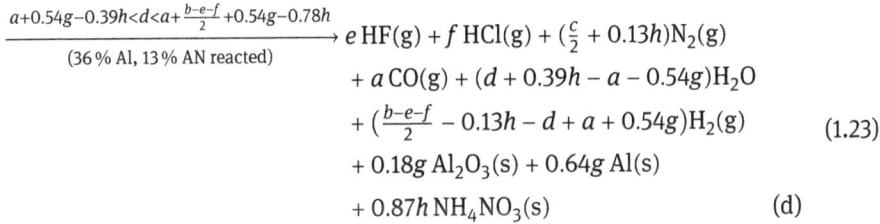

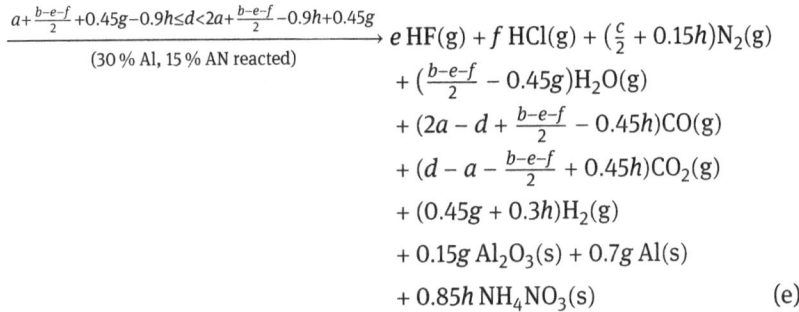

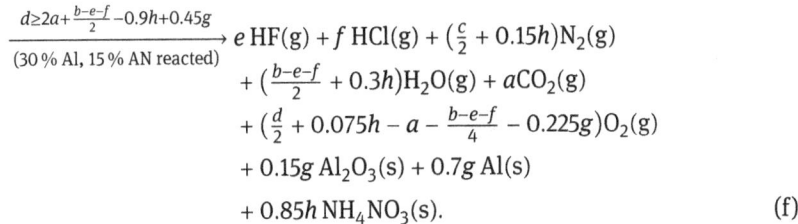

For example, HMX/Al(60/40) with formula $C_{0.812}H_{1.624}N_{1.624}O_{1.624}Al_{1.483}$ and $\Delta_f H^\theta(\text{HMX/Al}(60/40)) = 15.19\,\text{kJ/mol}$ gives the following decomposition products and the heat of detonation using equation (1.5) as:

According to the general formula $C_aH_bN_cO_dF_eCl_fAl_g(NH_4NO_3)_h$, the decomposition path (d) in equation (1.23) is followed because

$$a + 0.54g - 0.39h < d < a + \frac{b - e - f}{2} + 0.54g - 0.78h$$

$$\rightarrow 0.812 + 0.54 \times 1.483 < 1.624 < 0.812 + \frac{1.624}{2} + 0.54 \times 1.483$$

$$\rightarrow 1.612 < 1.624 < 2.425,$$

$$C_{0.812}H_{1.624}N_{1.624}O_{1.624}Al_{1.483}$$
$$\rightarrow 0.812N_2(g) + 0.812CO(g) + 0.011H_2O + 0.801H_2(g) + 0.267Al_2O_3(s) + 0.949Al(s),$$

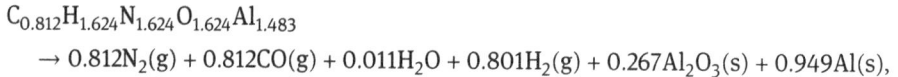

$$Q_{det}[H_2O(g)] \cong -\frac{1}{\text{Formula weight of HMX/Al(60/40)}}[0.011\Delta_fH^\theta(H_2O(g))$$
$$+ 0.812\Delta_fH^\theta(CO) + 0.267\Delta_fH^\theta(Al_2O_3) - \Delta_fH^\theta(HMX/Al(60/40))]$$
$$= -\frac{1}{100 \text{ g/mol}}[(0.011)(-241.8 \text{ kJ/mol})$$
$$+ (0.812)(-110.5 \text{ kJ/mol}) + (0.267)(-1675.7 \text{ kJ/mol}) - (15.19 \text{ kJ/mol})]$$
$$= 5.55 \text{ kJ/g.}$$

Summary

The theoretical computation of new energetic materials allows the identification of promising candidates for additional study and elimination of poor candidates. In the recent past, some theoretical methods have been used to predict the Q_{expl} of organic energetic compounds because the detonation behavior of explosives has received great interest. Since the Q_{expl} is a convenient parameter for describing the performance potential of energetic materials, different approaches have been described and demonstrated in this chapter.

Questions and problems

i The necessary information for some problems are given in the Appendix.

(1) What is the relationship between the heat of detonation and the condensed phase heat of formation of an explosive?

(2) What are the values of $Q_{det}[H_2O(l)]$ and $Q_{det}[H_2O(g)]$ on the basis of the decomposition paths given in equation (1.11) for PBX-9010?

(3) If the gas phase heats of formation of ethyl nitrate and TATB are −54.37 and 19.4 kcal/mol respectively, calculate $Q_{det}[H_2O(l)]$ for these energetic compounds on the basis of equations (1.15) and (1.16).

(4) The solid-phase heat of formation of styphnic acid $(C_6H_3N_3O_8)$ is −523.0 kJ/mol [1].
   (a) Calculate $Q_{det}[H_2O(l)]$ of this compound from equation (1.17).
   (b) Calculate $Q_{det}[H_2O(l)]$ of this compound from the decomposition paths outlined in equations (1.10) and (1.11).
   (c) If the measured value of $Q_{det}[H_2O(l)]$ is 2.952 kJ/g [1], arrange the reliability of the predicted results by the mentioned methods.

(5) Use equations (1.18) or (1.19) to calculate $Q_{det}[H_2O(l)]$ for the following explosives and compare the predicted results with the corresponding experimental values:
   (a) 3-Nitro-1,2,4-triazole-5-one (NTO) $(\Delta_f H^\theta(c) = -14.3\,kcal/mol$ [1];
       Exp. = 3.148 kJ/g [1]);
   (b) ε-Hexaazaisowurtzitane (CL-20) $(\Delta_f H^\theta(c) = 90.2\,kcal/mol$ [1];
       Exp. = 6.314 kJ/g [1]);
   (c) 1,1-Diamino-2,2-dinitroethylene (FOX-7) $(\Delta_f H^\theta(c) = -32.0\,kcal/mol$ [1];
       Exp. = 4.755 kJ/g [1];
   (d) Octanitrocubane (ONC) $(\Delta_f H^\theta(c) = 144.0\,kcal/mol$ [1]; Exp. = 7.648 kJ/g [1]).

(6) For Manitol hexanitrate $(C_6H_8N_6O_{18})$, calculate $Q_{det}[H_2O(l)]$ using equation (1.20).

(7) Picramic acid has the following molecular structure. Calculate the value of $Q_{det}[H_2O(l)]$ for this compound on the basis of equation (1.21).

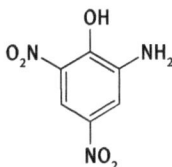

(8) Rice and Hare [15] have predicted $Q_{det}[H_2O(l)]$ using quantum mechanical calculations for NTO, CL-20, and FOX-7 as 4.711, 6.945, and 5.971, respectively. Use equation (1.22) to calculate $Q_{det}[H_2O(l)]$ for these compounds. According to the measured values of these compounds given in Problem (5), compare the calculated results of equation (1.22) with the corresponding results of Rice and Hare [15].

(9) ALEX32 with composition 37.4/27.8/30.8/4 RDX/TNT/Al/Wax and chemical formula $C_{1.65}H_{2.19}N_{1.38}O_{1.74}Al_{1.14}$ has $\Delta_f H^\theta(\text{ALEX32}) = -5.16$ kJ/mol. Find the heat of detonation of ALEX32.

For answers and solutions, please see p. 119

i

2 Detonation temperature

2.1 Adiabatic combustion (flame) temperature

2.1.1 Combustion of fuels with air

For a combustion process, in the absence of any work interactions and any changes in kinetic or potential energies, the stored chemical energy released is either lost as heat to the surroundings or is used internally to raise the temperature of the combustion products. The smaller the heat loss, the larger is the temperature rise that can be expected. If there is no heat loss to the surroundings, i. e. a combustion process that takes place adiabatically, the temperature of the products will reach a maximum, which is called the **adiabatic flame** or **adiabatic combustion temperature** of the reaction (Figure 2.1).

For a combustion process under conditions of constant volume and constant pressure, there are two types of adiabatic flame temperature describing the temperature of the combustion products theoretically reach if no energy is lost to the outside environment. The **constant volume adiabatic flame temperature** shows maximum temperature from a complete combustion process that occurs without any work (W), heat transfer (Q), or changes in kinetic or potential energy. To obtain the constant volume adiabatic flame temperature thermodynamically for closed reaction systems, the following equation for no heat transfer ($Q = 0$) and no work ($W = 0$) can be used:

$$U_{\text{products}} = U_{\text{reactants}}, \tag{2.1}$$

where U_{products} represents the internal energy of the products and $U_{\text{reactants}}$ corresponds to the internal energy of the reactants. The **constant pressure adiabatic flame temperature** indicates the maximum value for the temperature from a complete combustion process that occurs without any heat transfer or changes in the kinetic or potential energy. The constant pressure adiabatic flame temperature can also be calculated for steady-flow systems under conditions $Q = 0$ and $W = 0$ since

$$H_{\text{products}} = H_{\text{reactants}}, \tag{2.2}$$

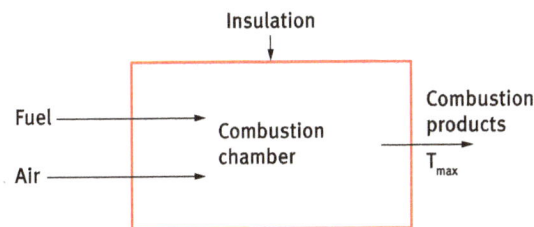

Figure 2.1: Maximum temperature (T_{max}) when combustion is complete and there is no heat loss.

https://doi.org/10.1515/9783110677652-002

where $H_{products}$ corresponds to the enthalpy of the products and $H_{reactants}$ is the enthalpy of the reactants.

Since there is no change in the volume of the system (i. e. generated work), the constant volume adiabatic flame temperature is higher than the constant pressure adiabatic flame temperature. For heat transfer, incomplete combustion, and dissociation, the temperature in the combustion chamber is lower than the adiabatic flame temperature (Figure 2.2).

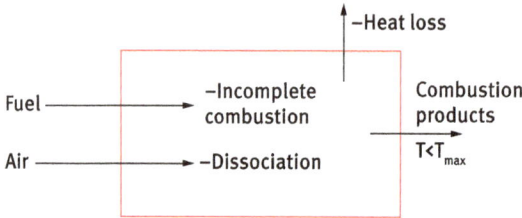

Figure 2.2: The temperature of the combustion products (T) is lower than T_{max} due to incomplete combustion, dissociation and heat loss.

The minimum amount of air needed for the complete combustion of a fuel, whereby all of the carbon in the fuel burns to CO_2, all the hydrogen burns to H_2O, and all the sulfur (if any) burns to SO_2, is called the **stoichiometric** or **theoretical** air. The value of the adiabatic flame temperature of a fuel is not unique and depends on the state of the reactants, the degree of completion of the reaction, and the amount of air used. The adiabatic flame temperature attains its maximum value when complete combustion occurs with the theoretical amount of air.

In order to calculate the constant volume and constant pressure adiabatic flame temperature through equations (2.1) and (2.2) for the combustion of solid or liquid reactants at standard state conditions, equation (1.4) for the determination of $H_{reactants}$ or $U_{reactants}$ can be used. Since the temperature of the products is not known prior to the calculations, the calculation of the enthalpy or internal energy of the products $H_{products}$ or $U_{products}$ is not so straightforward. Thus, the determination of the adiabatic flame temperature usually requires the use of an iterative technique. In such a technique, the values of $H_{products}$ or $U_{products}$ are determined for several high temperatures until the value of $H_{products}$ or $U_{products}$ becomes equal or close to $H_{reactants}$ or $U_{reactants}$. The constant volume and constant pressure adiabatic flame temperature is then determined from these two results by interpolation. Using air as the oxidant, the product gases mostly consist of N_2, and a good first guess for the adiabatic flame temperature is obtained by treating the entire product gases as N_2. Since the highest temperature to which a material can be exposed is limited by metallurgical considerations, the constant volume and constant pressure adiabatic flame temperature is an important consideration in the design of combustion chambers, gas turbines, and nozzles. The measured maximum temperatures are considerably lower than the constant volume and pressure adiabatic flame, because combustion is usually incomplete, some heat loss takes place, and some combustion gases dissociate at high temperatures (Figure 2.2). To lower the maximum temperature in a combustion chamber, excess air can be used to serve as a coolant. For a given fuel and oxidizer combination in the correct stoichiometric mixture, (i. e. correct proportions such that all fuel and all oxidizer are consumed), the amount of excess air can be tailored as part of the design to control the adiabatic flame temperature.

2.1.2 Combustion of propellants

For the adiabatic combustion of a desired propellant at a constant-pressure state, the enthalpy of a system should be constant during the process, i. e. the condition of equation (2.2) must be satisfied. Due to the existence of complex calculations for different categories of propellants, different computer codes such as ISPBKW [12], CHEETAH [11], and ZMWNI [38] can be used. In various numerical computer codes, calculations aim to seek a minimum for the thermodynamic potential for an assigned pressure and enthalpy. The constant pressure adiabatic flame temperature, (the so-called **adiabatic combustion temperature**), is determined on the basis of the weight percent of the different constituents. For example, Grys and Trzciński [38] compare the results of equilibrium calculations for exemplary mixtures containing polytetrafluoroethylene (PTFE) and magnesium powder at a pressure of 1 atm (Table 2.1). They used the equation of state BKWS database [39] for compounds containing fluorine and magnesium. It should be mentioned that BKW-EOS (equation (1.13)) has three different parameterizations including BKWC-EOS, BKWR-EOS, and BKWS-EOS. In contrast to the BKWR-EOS and BKWC-EOS, the covolumes used in the BKWS-EOS are assumed to be invariant and are based on the molecular structure of the product species [39].

Table 2.1: Comparison of the product compositions obtained from the ZMWNI and CHEETAH codes for calculation of the constant pressure adiabatic combustion of a mixture containing 70 % PTFE and 30 % Mg (the composition of the products is given in mol of product per mol of explosive) [38].

	F_2Mg	F	CF_2	CF	FMg	CF_3
ZMWNI	5.64E−01	1.67E−01	1.75E−02	4.99E−02	6.75E−02	4.80E−05
CHEETAH	5.64E−01	1.67E−01	1.75E−02	4.99E−02	6.75E−02	4.80E−05
	CF_4	C_3	F_4Mg_2	Mg	C_2F_2	C_2
ZMWNI	1.80E−06	2.75E−02	3.30E−05	6.16E−03	3.76E−05	3.79E−03
CHEETAH	1.80E−06	2.75E−02	3.30E−05	6.16E−03	3.76E−05	3.79E−03
	F_2	C_5	C_2F_4	C_4	C_2F_6	*C solid
ZMWNI	1.52E−06	3.51E−04	1.52E−08	1.61E−04	9.93E−13	4.73E−01
CHEETAH	1.52E−06	3.51E−04	1.53E−08	1.61E−04	9.94E−13	4.73E−01

As is shown in Table 2.1, the differences between the equilibrium values calculated for the constant-pressure combustion using the CHEETAH and ZMWNI codes are small, below 0.2%. For solid energetic materials, the adiabatic combustion temperature is very often used in the numerical modeling of the combustion process. Since the mass percent of different constituents in a solid propellant can influence the values of the adiabatic combustion temperature, computer codes evaluate the equilibrium composition of various products as a function of the mass percent of different constituents of propellant at the specified pressure in the combustion chamber. Figure 2.3 shows the output of the option of the constant pressure combustion using the ZMWNI code for mixtures of PTFE with Mg or Al, which enables the adiabatic combustion temperature on the basis of the mass fraction of a metal to be determined.

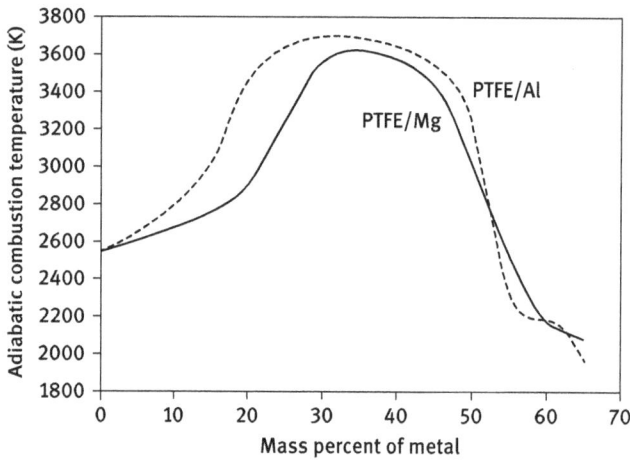

Figure 2.3: Dependence of the adiabatic combustion temperature on the mass percent of the metal in PTFE/Al and PTFE/Mg mixtures [38].

2.2 Detonation (explosion) temperature for explosives

2.2.1 Measurement of detonation temperature

The detonation process for an explosive is extremely fast, meaning that the gases do not have time to expand to any great extent. As seen in Chapter 1, the heat of detonation raises the temperature of the gases, which in turn causes the gases to expand and work on the surroundings. Therefore, as is the case for the adiabatic combustion temperature or flame temperature, it can be assumed that the detonation products can attain the maximum temperature under adiabatic conditions. The effect of the transfer of heat energy on the gaseous detonation products can be used to calculate the detonation temperature or explosion temperature.

Determination of the detonation temperature is extremely important, because it can be related to the study of the kinetics of chemical reactions in the reaction zone and the thermodynamic state of the detonation products. The detonation temperature is the most important thermodynamic quantity needed to analyze the kinetics of explosion transformation and to evaluate the correctness of the equations of state (EOS) which are required to describe explosive processes. Optical methods are frequently used to determine the temperature at the detonation wave front and the temperature of the detonation products. To measure the temperature in shock and detonation waves with submicrosecond time resolution, photoelectric recording of the light intensity in high temperature phenomena is used [40]. It should be mentioned that there are numerous different measuring systems for the determination of the detonation temperature, which mainly differ with respect to the optical system which is used. For example, Sil'vestrov et al. used an optical pyrometer to measure the brightness temperature of the detonation front of an emulsion explosive with glass microballoons as a sensitizer [41].

The brightness of the detonation front interacting with a detector which can be used to measure the detonation temperature has an absolute accuracy estimated to be ±100 K for liquid explosives and ±200 K for solid explosives. Any voids or density discontinuities can lead to measurement of the brightness of the shocked air or shocked detonation products rather than the Chapman–Jouguet (C–J) detonation products. Thus, a density discontinuity-free system such as a liquid or a single crystal is useful for measurement of the detonation temperature. Since measurement of the detonation temperature is difficult, experimental data for detonation temperatures is scarce. For reporting detonation temperatures, a blackbody of equivalence photographic brightness with an absolute accuracy of about 200 K is usually used [12].

For measurements of the detonation temperature of condensed heterogeneous opaque (or partially transparent) explosives, there is a large scatter of results. This situation is due to the different qualities of the prepared explosive charges, as well as additional glow effects at the interface between the explosive and optically transparent medium [41]. For many heterogeneous explosives – from low-velocity safety explosives, low-density TNT, and PETN to pressed RDX and HMX – despite the methodological difficulties, the brightness temperatures of the detonation front are found in the range 2350–7500 K [40, 42]. The reported detonation temperatures for a specific explosive depend on the method which is used. For example, the measured detonation temperature for PETN at a density of $\approx 1.6\,g/cm^3$ is 4250–6300 K [41]. If an EOS corresponding to experimental data is constructed, the temperature measurements are in satisfactory agreement with the results of calculations for different EOSs for the explosion products.

2.2.2 Calculation of detonation temperature

2.2.2.1 Computer code

Many thermochemical codes are currently being used to carry out thermodynamic calculations of the detonation temperature of the condensed explosives, e. g. BKW Fortran [12], CHEETAH [11], and EXPLO5 [10]. For a constant volume explosion, conservation of the internal energy is a physical condition, and the temperature is an unknown state parameter under this condition. The main aim in thermochemical codes is to determine the composition of the products for which the principle of the conservation of the internal energy is fulfilled and the thermodynamic potential reaches its minimum. Among different EOSs that can be used in these thermochemical codes, it was found that the BKWS-EOS predictions of the detonation temperature are relatively good for $C_aH_bN_cO_d$ explosives. It was shown that the overall percent RMS error in the predicted detonation temperatures of $C_aH_bN_cO_d$ explosives is higher for the JCZS-EOS and BKWR-EOS than for the BKWS-EOS. Sućeska uses EXPLO5 with BKW-EOS for the gaseous detonation products and Cowan–Fickett's equation of state for solid carbon to calculate the detonation parameters at the CJ point [10]. He compared the values of the detonation temperature at the CJ point calculated by EXPLO5 to the measured values, which are given in Figure 2.4. As is shown in Figure 2.4, a good agreement exists between them (about 5 %). In other research work, it was found that the values for the detonation temperature which were calculated by applying the BKWR-EOS set of constants are lower than those obtained experimentally, with a mean difference of about 550 K (about 20 %) [43].

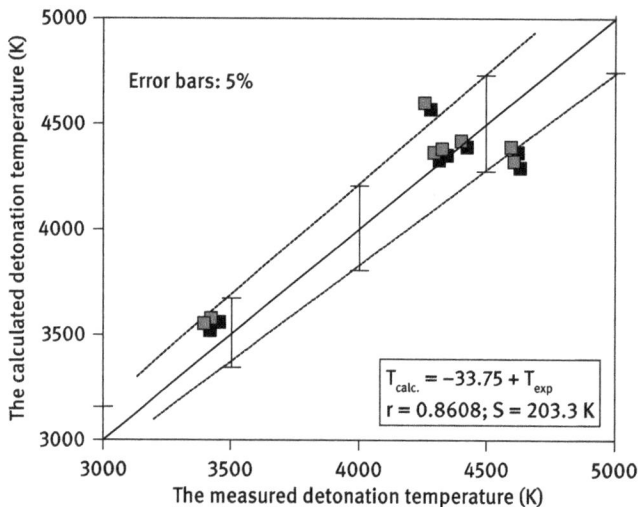

$$T_{calc.} = -33.75 + T_{exp}$$
$$r = 0.8608; S = 203.3 K$$

Figure 2.4: Comparison of the calculated and measured values of the detonation temperature for several $C_aH_bN_cO_d$ explosives by EXPLO5 using BKW-EOS [10].

2.2.2.2 The use of heat capacity

For calculation of the detonation temperature in this approach, it is assumed that the explosive at an initial temperature T_i is converted to gaseous products which are also at T_i. The temperature of these gaseous products is then raised to the detonation (explosion) temperature T_{det} by Q'_{det}, where

$$Q'_{det} = \sum n_j \Delta_f H^\theta (\text{detonation product})_j - \Delta_f H^\theta (\text{explosive}).$$

Thus, the value of T_{det} depends on the value of Q_{det} and on the separate molar heat capacities of the gaseous products as follows:

$$Q'_{det} = \int_{T_i}^{T_{det}} \sum n_j \bar{C}_V (\text{detonation product})_j \, dT, \tag{2.3}$$

where $\bar{C}_V(\text{detonation product})_j$ and n_j are the molar heat capacity at constant volume and the number of mole of the j-th detonation product, respectively. In using this equation, several points should be considered.

(1) The values of $\bar{C}_V(\text{detonation product})_j$ for various gaseous products are a function of temperature. Since the heat capacities of the gaseous products vary with temperature in a nonlinear manner, there is no simple relationship between temperature and $\bar{C}_V(\text{detonation product})_j$. Thus, the tabulated mean molar heat capacities of gaseous products at various temperatures can be used for this purpose. The rise in temperature of the gaseous products is calculated by dividing the heat generated Q'_{det} by the mean molar heat capacity of the gases at constant volume according to

$$T_{det} = \frac{Q'_{det}}{\sum n_j \bar{C}_V (\text{detonation product})_j} + T_i. \tag{2.4}$$

(2) Since the molar heat capacities at constant pressure as a function for gaseous products are more readily available in the literature (e. g. from the NIST Chemistry WebBook) than $\bar{C}_V(\text{detonation product})_j$, the approximate detonation temperature can be obtained by

$$T_{det,app} = \frac{Q'_{det}}{\sum n_j \bar{C}_P (\text{detonation product})_j} + T_i, \tag{2.5}$$

where $\bar{C}_P(\text{detonation product})_j$ are the molar heat capacities of the products at constant pressure.

(3) When using the heat capacity method, it is essential to know the detonation products. Suitable which can be used for this purpose methods are given in Section 1.2.

2.2.2.3 Empirical methods

Computer codes require expert users and appropriate EOSs for calculation of the detonation temperature, as well as the availability of the desired computer code. Moreover, the use of the heat capacity has some limitations, such as knowledge of the decomposition products and the ability to write a computer program on the basis of the heat capacity of the products as a function of temperature. Several empirical methods have also been introduced to evaluate the detonation temperature of $C_a H_b N_c O_d$ high explosives on the basis of their heat of formation and molecular structures.

The condensed phase heat of formation

It was shown that on the basis of the decomposition products of equation (1.11), the detonation temperature of a high explosive with the general formula $C_a H_b N_c O_d$ is obtained on the basis of the following four conditions [44]:

$$T_{det} = T_i + \begin{cases} \dfrac{\Delta_f H^\theta(\text{explosive}) - 529.4d}{0.01095a - 0.1132b + 0.01335c - 0.09910d} \\ \quad \text{with } 0 \le a - d \qquad\qquad\qquad\qquad\qquad\qquad\quad (a) \\[2mm] \dfrac{\Delta_f H^\theta(\text{explosive}) - 943.4a + 1230d}{-0.1914a + 0.05967b + 0.01687c + 0.2224d} \\ \quad \text{with } 0 > a - d \text{ and } 0 < \frac{b}{2} - d + a \qquad\qquad (b) \\[2mm] \dfrac{\Delta_f H^\theta(\text{explosive}) - 172.46a - 20.58b + 283.0d}{0.01219a + 0.01584b + 0.01866c + 0.02530d} \\ \quad \text{with } d - a - \frac{b}{2} \ge 0 \text{ and } 0 \le 2a - d + \frac{b}{2} \quad (c) \\[2mm] \dfrac{\Delta_f H^\theta(\text{explosive}) + 625.2a - 142.8b}{0.05905a - 0.04381b + 0.01866c + 0.02036d} \\ \quad \text{with } 0 > 2a - d + \frac{b}{2} \qquad\qquad\qquad\qquad\quad (d) \end{cases} \qquad (2.6)$$

For example, let us consider the use of equation (2.6) for the calculation of PETN with formula $C_5 H_8 N_4 O_{12}$ and $\Delta_f H^\theta(\text{PETN}) = -538.48$ kJ/mol (Appendix). Since the condition of equation (2.6) (c) is satisfied for PETN, T_{det} is calculated as

$$T_{det} = T_i + \frac{\Delta_f H^\theta(\text{explosive}) - 172.46a - 20.58b + 283.0d}{0.01219a + 0.01584b + 0.01866c + 0.02530d}$$

$$= 298 + \frac{74.8 - 172.46(5) - 20.58(8) + 283.0(12)}{0.01219(5) + 0.01584(8) + 0.01866(4) + 0.02530(12)}$$

$$= 3532 \text{ K}.$$

Gas phase heat of formation

As was shown in Section 1.3.1.1, the linear combination of the elemental composition of the explosive and the estimated standard gas phase heat of formation of the explosive, $\Delta_f H^\theta(g)$ can be used to derive two reliable correlations for obtaining the heats of detonation of aromatic and nonaromatic explosives. It was shown that crystal effects can also be excluded for determining the detonation temperature in this manner [45, 46] because the crystalline heat of formation can be correlated with the gas phase heat of formation for some classes of explosives [47]. Since it can be stated very approximately that the detonation temperature is proportional to the heat of detonation, the results indicated that the following equations can be used to predict the detonation temperature of aromatic and nonaromatic explosives [48]:

$$(T_{det})_{aromatic} = \left(-75.8 + \frac{\left(\begin{array}{c}950.8a + 12.3b + 1114.9c + 1324.5d \\ + 0.287[\Delta_f H^{\theta}(g) \text{ of explosive}]\end{array}\right)}{\begin{array}{c}\text{formula weight} \\ \text{of explosive}\end{array}} \right) \times 10^3, \quad (2.7)$$

$$(T_{det})_{nonaromatic} = \left(149.0 + \frac{\left(\begin{array}{c}-1513.9a - 196.5b - 2066.1c - 2346.2d \\ + 0.287[\Delta_f H^{\theta}(g) \text{ of explosive}]\end{array}\right)}{\begin{array}{c}\text{formula weight} \\ \text{of explosive}\end{array}} \right) \times 10^3. \quad (2.8)$$

These equations provide a simple method for the estimation of the detonation temperature of pure explosives, which require as input information only the elemental composition and heat of formation of the explosive in the gas phase, which can be calculated using quantum mechanical, group additivity, and empirical methods [17].

To demonstrate the application of this method for aromatic and nonaromatic high explosives, the explosives Hexanitrohexaazaisowurtzitane (HNIW or CL 20) and hexanitrostilbene (HNS) with the formulas $C_6H_6N_{12}O_{12}$ and $C_{14}H_6N_6O_{12}$, respectively, will be considered here. As was given in Section 1.3.1.1, the predicted $\Delta_f H^{\theta}(g)$ values for CL 20 and HNS are 500 and 150 kJ/mol, respectively. Thus, the use of these values in equations (2.7) and (2.8) gives the following detonation temperatures for the aromatic and non-aromatic explosives, HNS and CL-20 respectively:

$$(T_{det})_{aromatic} = \left(-75.8 + \frac{\left(\begin{array}{c}950.8(14) + 12.3(6) + 1114.9(6) \\ +1324.5(12) + 0.287(150)\end{array}\right)}{450.23} \right) \times 10^3$$

$$= 4185 \text{ K},$$

$$(T_{det})_{nonaromatic} = \left(149.0 + \frac{\left(\begin{array}{c}-1513.9(6) - 196.5(6) - 2066.1(12) \\ -2346.2(12) + 0.287(500)\end{array}\right)}{438.18} \right) \times 10^3$$

$$= 5072 \text{ K}.$$

The predicted detonation temperatures for HNS and CL 20 are consistent with the corresponding values of $Q_{det}[H_2O(l)]$ given in Section 1.3.1.1.

Structural parameters

All of the chemical bonds present in the reacting molecules are broken in the detonation process and the reactive species recombine to form stable products. The ratio of oxygen to carbon and hydrogen is important for obtaining a high or low detonation temperature. The presence of some structural parameters, such as specific functional groups, may also affect the value of the detonation temperature. It was shown that the following equation can supply a suitable pathway for predicting the detonation temperatures of high explosives with the general formula $C_aH_bN_cO_d$ [49]:

$$T_{det} = 5136 - 190.1a - 56.4b + 115.9c + 148.4d - 466.0(d/a)$$
$$- 700.8(b/d) - 282.9n_{NH_x}, \qquad (2.9)$$

where n_{NH_x} is the number of $-NH_2$ and NH_4^+ moieties in the energetic compounds. This equation provides the simplest empirical procedure for the estimation of the detonation temperature. For example, application of equation (2.9) for TACOT (2,4,8,10-Tetranitro-5H-benzotriazolo[2,1,a]-benzotriazol-6-ium, hydroxide, inner salt) with formula $C_{12}H_4N_8O_8$ is given as follows:

$$T_{det} = 5136 - 190.1(12) - 56.4(4) + 115.9(8) + 148.4(8) - 466.0(8/12)$$
$$- 700.8(4/8) - 282.9(0)$$
$$= 4083\,K.$$

The predicted detonation temperature for TACOT using a computer code with BKWR-EOS and BKWS-EOS is 3330 and 4040 K, respectively. As was previously mentioned, the BKWS-EOS gives better predictions for detonation temperature of high explosives. Therefore, the calculated detonation temperatures using this empirical method are close to those obtained by complex thermochemical codes using appropriate equations of state.

2.2.2.4 A simple pathway for the prediction of the detonation temperature of a mixture of explosives

Since explosives can be used in mixed compositions in order to optimize their performance, it is possible to use the data for the pure explosives to estimate the detonation temperature for a mixture of explosives. It was shown that the following equation is the simplest way to obtain acceptable results [48]:

$$T_{det,mix} = \sum_j x_j T_{det,j}, \qquad (2.10)$$

where x_j is the mole fraction of the j-th component in the mixture of explosives. For example, pentolite-50/50 comprises a mixture of 50 % PETN and 50 % TNT ($C_{2.33}H_{2.37}N_{1.29}O_{3.22}$). The calculated $\Delta_f H^\theta(g)$ values for TNT ($C_7H_5N_3O_6$) and PETN ($C_5H_8N_4O_{12}$) are 3 and −457.7 kJ/mol from the NIST Chemistry WebBook [8] in which Benson's group additivity method can be used [19], respectively. Therefore, the value of the detonation temperature for this binary mixture can be obtained according to

$$(T_{det})_{aromatic} = \left(-75.8 + \frac{\begin{pmatrix} 950.8(7) + 12.3(5) + 1114.9(3) \\ + 1324.5(6) + 0.287(3) \end{pmatrix}}{227.13} \right) \times 10^3$$

$$= 3492\,K,$$

$$(T_{det})_{nonaromatic} = \left(149.0 + \frac{\begin{pmatrix} -1513.9(5) - 196.5(8) - 2066.1(4) \\ - 2346.2(12) + 0.287(-457.7) \end{pmatrix}}{316.14} \right) \times 10^3$$

$$= 4470\,K,$$

$$T_{det,mix} = x_{TNT}(T_{det})_{TNT} + x_{PETN}(T_{det})_{PETN}$$
$$= \left(\frac{\frac{50}{227.13}}{\frac{50}{227.13} + \frac{50}{316.14}} \right)(3492) + \left(\frac{\frac{50}{316.14}}{\frac{50}{227.13} + \frac{50}{316.14}} \right)(4470) = 3901\,K.$$

The calculated detonation temperature of PENTOLITE using a computer code employing the BKWR-EOS and BKWS-EOS is 3360 and 4030 K, respectively. As was mentioned before, BKWS-EOS gives better predictions for the detonation temperature of high explosives. Thus, the calculated detonation temperatures using this empirical method are close to those obtained using complex thermochemical codes using the appropriate equation of state.

2.2.3 Calculation of the detonation temperature of ideal and non-ideal energetic compounds

As seen in previous sections, some attempts have been done for simple evaluation or desk calculation of the detonation temperature of ideal and less ideal pure or a mixture of CHNO explosives [48, 49]. A simple model has been introduced recently for calculating the detonation temperature of important classes of ideal, less ideal and non-ideal energetic compounds [37]. It can be used for wide classes of energetic materials including ideal pure explosives or energetic mixtures with the general formula of CHNOFCl as well as plastic bonded explosives (PBXs) and non-ideal composite explosives containing Al or AN. The model provides good prediction of the detonation temperatures as compared with the outputs of computer codes using appropriate equations of state, e. g. the computed results of BKWR-EOS and BKWS-EOS for various ideal and non-ideal explosives [37]. It is based on a linear relationship between the detonation temperature and the heat of detonation, using equations (1.5) and (1.23), of the following form [37]:

$$T_{det} = 1883 + 522Q_{det}, \tag{2.11}$$

where T_{det} and Q_{det} are in K and kJ/g, respectively. For example, TNT/Al (89.4/10.6) with formula $C_{2.756}H_{1.969}N_{1.181}O_{2.362}Al_{0.393}$ and $\Delta_f H^\theta$(TNT/Al (89.4/10.6)) = –24.73 kJ/mol (see Appendix) gives the following decomposition products and the heat of detonation using equation (1.5) as:

According to the general formula $C_a H_b N_c O_d F_e Cl_f Al_g (NH_4NO_3)_h$, the decomposition path (d) in equation (1.23) is followed because

$$d \leq a + 0.375g - 0.3h \rightarrow 2.362 \leq 2.756 + 0.375 \times 0.393 \rightarrow 2.362 \leq 2.903,$$

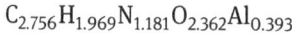

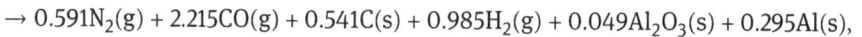

$C_{2.756}H_{1.969}N_{1.181}O_{2.362}Al_{0.393}$
$\rightarrow 0.591N_2(g) + 2.215CO(g) + 0.541C(s) + 0.985H_2(g) + 0.049Al_2O_3(s) + 0.295Al(s),$

$$Q_{det}[H_2O(g)] \cong \frac{-[2.215\Delta_f H^\theta(CO) + 0.049\Delta_f H^\theta(Al_2O_3) - \Delta_f H^\theta(TNT/Al(89.4/10.6))]}{\text{Formula weight of TNT/Al(89.4/10.6)}}$$
$$= \frac{-[(2.215)(-110.5\,kJ/mol) + (0.049)(-1675.7\,kJ/mol) - (-24.73\,kJ/mol)]}{100\,g/mol}$$
$$= 3.02\,kJ/g,$$

$$T_{det} = 1883 + 522Q_{det} = 1883 + 522 \times 3.02 = 3459\,K.$$

The calculated detonation temperature with BKWS-EOS using the partial equilibrium is 3910 K [39].

Due to large deviations between the calculated results of equation (2.11) and those computed by one of the best available equations of state, i. e. BKWS-EOS using the partial equilibrium, for AN based and liquid explosives, equation (2.11) should be corrected. It can be expected that part of Q_{Det} is used to several phase changes in the solid state of AN [50]. Thus, the calculated value of T_{det} from equation (2.11) should be corrected for AN-based explosives as [37]:

$$T_{det,corr\,AN} = (-0.88 \times \%AN/100 + 1.22) \times T_{det}, \tag{2.12}$$

where $T_{det,corr\,AN}$ is the corrected detonation temperature of AN-based explosives in K; and %AN is the percentage of ammonium nitrate in AN-based explosives. For example, AMATOL-80/20 corresponding to 80/20 AN/TNT has the formula $C_{0.62}H_{0.44}N_{0.26}O_{0.53}(AN)_1$ and $\Delta_f H^\theta$(AMATOL-80/20) = –371.25 kJ/mol (see Appendix) gives the following decomposition products and the heat of detonation using equation (1.5) as:

The decomposition path (e) in equation (1.23) is followed because

$$a + \frac{b-e-f}{2} + 0.45g - 0.9h \le d < 2a + \frac{b-e-f}{2} - 0.9h + 0.45h$$

$$\rightarrow 0.62 + \frac{0.44}{2} - 0.9 \le 0.53 < 2 \times 0.62 + \frac{0.44}{2} - 0.9 \rightarrow -0.06 \le 0.53 < 0.56,$$

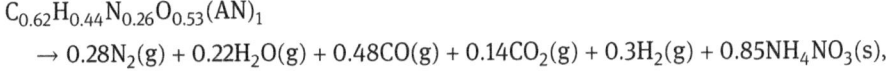

$$C_{0.62}H_{0.44}N_{0.26}O_{0.53}(AN)_1$$

$$\rightarrow 0.28N_2(g) + 0.22H_2O(g) + 0.48CO(g) + 0.14CO_2(g) + 0.3H_2(g) + 0.85NH_4NO_3(s),$$

$$Q_{det}[H_2O(g)] \cong -\frac{1}{\text{Formula weight of } C_{0.62}H_{0.44}N_{0.26}O_{0.53}(AN)_1}$$
$$\times [0.22\Delta_f H^\theta(H_2O(g)) + 0.48\Delta_f H^\theta(CO) + 0.14\Delta_f H^\theta(CO_2)$$
$$+ 0.85\Delta_f H^\theta(NH_4NO_3) - \Delta_f H^\theta(C_{0.62}H_{0.44}N_{0.26}O_{0.53}(AN)_1)]$$

$$= -\frac{1}{100 \text{ g/mol}}[(0.22)(-241.8 \text{ kJ/mol}) + (0.48)(-110.5 \text{ kJ/mol})$$
$$+ (0.14)(-393.5 \text{ kJ/mol}) + (0.85)(-365.1 \text{ kJ/mol}) - (-371.25 \text{ kJ/mol})]$$

$$= 1.00 \text{ kJ/g},$$

$$T_{det} = 1883 + 522 Q_{det} = 1883 + 522 \times 1.00 = 2405 \text{ K},$$

$$T_{det,corr \text{ AN}} = (-0.88 \times \%AN/100 + 1.22) \times T_{det}$$
$$= (-0.88 \times 0.80 + 1.22) \times 2405 = 1241 \text{ K}.$$

The calculated detonation temperature with BKWS-EOS using the partial equilibrium is 1440 K [39].

Since the calculated detonation temperature of liquid explosives such as nitroglycerine by equation (2.11) is lower than that of solid explosives, equation (2.11) was corrected for liquid explosives as [37]:

$$T_{det,corr \text{ liquid}} = 0.87 \times T_{det}, \tag{2.13}$$

where $T_{det,corr \text{ liquid}}$ is the corrected detonation temperature of liquid explosives in K. For example, tetranitromethane has $\Delta_f H^\theta(TNM) = 38$ kJ/mol (see Appendix) gives the following decomposition products and the heat of detonation using equation (1.5) as:

The decomposition path (f) in equation (1.23) is followed because

$$d \ge 2a + \frac{b-e-f}{2} - 0.9h + 0.45g \rightarrow 8 \ge 2,$$

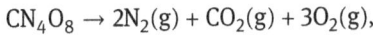

$$CN_4O_8 \rightarrow 2N_2(g) + CO_2(g) + 3O_2(g),$$

$$Q_{det}[H_2O(g)] \cong \frac{-[\Delta_f H^\theta(CO_2) - \Delta_f H^\theta(TNM)]}{\text{Formula weight of TNM}}$$
$$= \frac{-[(1)(-393.5 \text{ kJ/mol}) - (38\text{kJ/mol})]}{196.05 \text{ g/mol}} = 2.20 \text{ kJ/g},$$

$$T_{det} = 1883 + 522Q_{det} = 1883 + 522 \times 2.20 = 3031 \text{ K},$$

$$T_{det,corr \ liquid} = 0.87 \times T_{det} = 0.87 \times 3031 = 2637 \text{ K}.$$

The measured detonation temperature of TNM is 2800 K [39].

Summary

Different approaches have been reviewed for the calculation of the adiabatic flame temperature, combustion temperature, and detonation temperature for the combustion of fuel with air, deflagration of a propellant, and detonation of high explosives, respectively. For calculation of the detonation temperature, it is important to use appropriate EOS for the gaseous detonation products such as the BKWS-EOS in order to obtain reliable predictions. For $C_aH_bN_cO_d$ explosives, several empirical methods have been discussed for calculation of the detonation temperature. These empirical methods are the simplest methods for the evaluation of the detonation temperature.

Equations (2.11) to (2.13) also provide the simplest empirical procedure for estimation of the detonation temperatures of pure explosives or energetic mixtures with the general formula of CHNOFCl as well as PBXs and composite explosives containing Al or AN.

Questions and problems

i The necessary information for some problems are given in the Appendix.

(1) In designing a new explosive, how is it possible to increase its detonation temperature by considering the condensed phase heat of formation?
(2) Calculate the detonation temperature of COMP C-3 using equation (2.6).
(3) If the gas phase heat of formation of ABH is 47.9 kcal/mol, calculate its detonation temperature on the basis of equations (2.7) or (2.8).
(4) Calculate the detonation temperature of Cyclotol-70/30 using equation (2.9).
(5) AMATEX-40 with composition 42/20/38 AN/RDX/TNT and chemical formula $C_{1.44}H_{1.38}N_{1.04}O_{1.54}(AN)_{0.53}$ has $\Delta_f H^\theta(\text{AMATEX-40}) = -95.77$ kJ/mol. If $Q_{det}[H_2O(g)] = 2.34$ kJ/g, find the detonation temperature of AMATEX-40.

i For answers and solutions, please see p. 119.

3 Detonation velocity

During the detonation process, an explosive undergoes chemical reactions at very high speed which produces a shock wave or a detonation wave. As the chemical reaction is initiated instantaneously, high temperature and pressure gradients are generated at the wave front. The chemical reactions reinforce propagation of the detonation wave through the explosive. The **detonation velocity** is the velocity with which detonation waves travel through an explosive. It is the basic performance property which is a function of the energy produced by an explosive decomposition. The standard hydrodynamic theory for computing the detonation velocity of an explosive is independent of the chemical reactions which occur, and is concerned only with the amount of energy liberated and the nature of the end products.

The detonation velocity is a characteristic of each individual explosive and is not influenced by external factors if (i) the density of the explosive is at its maximum value and (ii) the explosive is charged into columns which are considerably wider than the critical diameter, i. e. the minimum diameter which is able to transmit the detonation. The value of the detonation velocity increases, with increasing density of packing of the explosive in the column.

3.1 Chapman–Jouguet (C–J) theory and detonation performance

During the dynamic action of the shock wave on the explosive, a thin layer of the explosive is compressed from the initial (loading) density, ρ_0, to higher density, ρ_1, in accordance with the shock (or **Hugoniot**) adiabatic curve for a given explosive. Under these conditions, the initial pressure, P_0, and temperature, T_0, increase to higher pressure, P_1, and temperature, T_1, in the compressed explosive layer where consequently initiation of the chemical reactions takes place. After completion of detonation reaction, the density, pressure and temperature reaches the values ρ_2, P_2, and T_2. This situation corresponds to the point lying on the shock adiabatic curve for the detonation products where they expand isentropically into the surrounding medium. For the **steady-state model of detonation**, it can be assumed that the points (ρ_0, P_0, T_0), (ρ_1, P_1, T_1), and (ρ_2, P_2, T_2) lie on one line that is called the **Rayleigh** or **Michelson** line. The detonation velocity of a given explosive can be used to determine the slope of the Rayleigh line. As shown in Figure 3.1, the **Chapman–Jouguet (C–J)** point corresponds to the end of the chemical reactions where the Rayleigh line is a tangent to the adiabatic shock of the detonation products at this point.

https://doi.org/10.1515/9783110677652-003

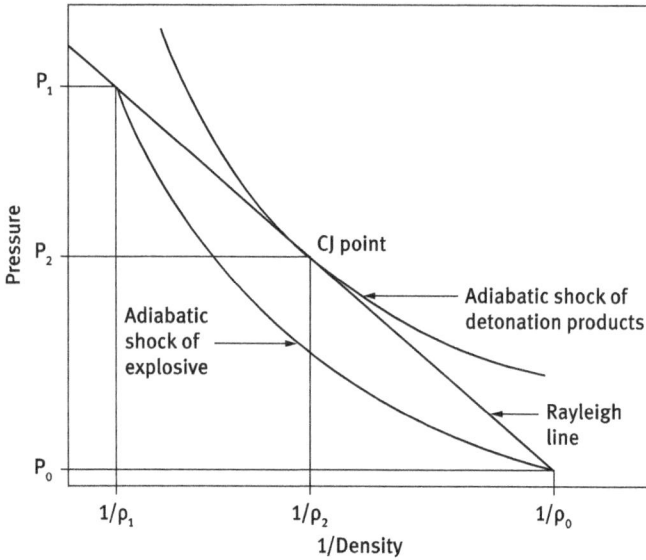

Figure 3.1: Adiabatic shock of the explosive and its detonation products in the case of steady detonation.

3.2 Ideal and nonideal explosives

A nonideal explosive has significantly different detonation properties than those predicted by some computer codes such as BKW [12], CHEETAH [11], and EXPLO5 [10] which use empirical equations of state such as BKW-EOS [39] and JCZ-EOS [51]. Two important characteristics of nonideal explosives include a high degree of inhomogeneity and the secondary exothermic reactions occurring in the detonation products expanding behind the detonation zone. For practical applications, ideal explosives should have short reaction zones and small failure diameters. Detonation velocities can typically be measured to within a few percent at various charge diameters. Since the measured results are a function of the charge diameters, the measured data are extrapolated to an infinite diameter for comparison with steady state calculations. The results of equilibrium, one-dimensional. and steady-state calculations are significantly different from the detonation properties of nonideal explosives. Since the amount of reacted explosive may be a function of the reaction zone length, physical separation of the fuel and oxidizer in nonideal explosives results in extended chemical reaction zones. Diffusion may also play a major role in the measured values of the detonation properties of nonideal explosives. For predicting the detonation properties of non-ideal explosives via a computer code, the partial equilibrium as a simple approximation can be used instead of a complex reaction mechanism.

Explosive nitrate salts and aluminized composite explosives are well-known examples of nonideal explosives. Ammonium nitrate (AN) and aluminum based explosives have been widely used as industrial and military explosives. AN is used in ANFO (ammonium nitrate and fuel oil), emulsion explosives and Amatol (a mixture of TNT and AN). Since their detonation velocities do not easily reach the theoretically predicted values given by well-known thermochemical computer codes: they show nonideal behavior. For most practical conditions these explosive will not achieve the ideal behavior which is predicted by thermodynamic theory. For AN based explosives, their nonideal behavior can be explained by the low decomposition rate of AN. This situation provides a wide reaction zone where the decomposition reactions can be extinguished by lateral heat loss and refraction waves. AN is one of the major components of most nitrate-based explosives and has a large value for the minimum diameter and relatively small value for the critical diameter. For most practical conditions, AN will never reach the ideal behavior which is predicted by hydrodynamic theory [39]. Due to either complete reaction or no reaction of AN with the reaction products, the differences found between the observed and calculated performance mean that AN based explosives can be classified as non-ideal explosives. If some percentage of AN is assumed to decompose and the rest remains intact, the measured detonation velocity can be reproduced by a thermochemical computer code [12]. For example, it was found that the experimental values for the detonation velocities of Amatex and Amatol can be reproduced by the BKW computer code, if 50 and 19 % of AN in them decomposes, respectively [12]. Various detonation temperatures can be assumed to be the cause of different amounts of AN decomposition, because at higher temperatures more AN decomposes in the explosive.

3.3 Measurement of the detonation velocity

Different methods can be used to determine the detonation velocity based on the measurement of the time interval needed for the detonation wave to travel a known distance through the explosive being tested. The simple method of **Dautriche** can be used to obtain a rough estimation of the detonation velocity [52] and does not require the use of any special and costly instruments. The accuracy of this method is relatively good and is within about 4.5 % [52].

As shown in Figure 3.2, the detonation reaction sets off the two arms (probes) of the detonating fuse which are embedded at a fixed distance in the cartridge. After initiation of the detonation, the detonating fuse has two waves traveling in the opposite direction. These two waves meet at a point, and the collision is marked on a lead or aluminum plate, where the following equation is used to measure the detonation velocity:

$$D_{det}(\text{explosive charge}) = D_{det}(\text{detonating fuse}) \times \frac{L}{2A}, \qquad (3.1)$$

where D_{det} (explosive charge) is the detonation velocity of the explosive being tested in m/s, D_{det} (detonating fuse) is the detonation velocity of the calibrated detonating

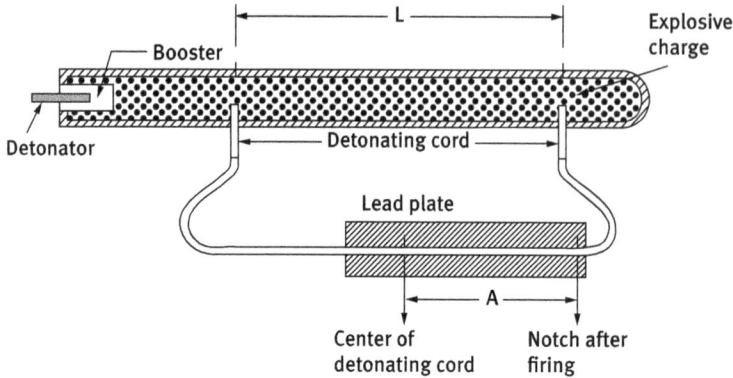

Figure 3.2: Dautriche method for measuring the detonation velocity.

fuse used in the test in m/s, A is the distance between the mark on the witness plate and center of the detonating fuse in cm, and L is the length between the probes in cm.

Optical and electrical methods are two conventional approaches used to determine the detonation velocity. Several conventional methods can be used to determine the detonation velocity, which include

(i)   optical methods,

(ii) electronic counter and velocity probes techniques,

(iii) oscilloscope and velocity probes technique,

(iv) probe for the continuous determination of the detonation velocity and oscilloscope technique,

(v) optical fibers as velocity probes.

Details of these methods have been reviewed elsewhere [52].

For explosives it is common to measure detonation velocities of the materials at nominal composition and density under ambient conditions in large charges [1]. To calculate detonation velocities under other conditions, specific equations have been developed as a function of the composition and density of the explosive, charge diameter, and temperature [9]. It should be mentioned that the critical diameter is the minimum diameter of an explosive charge at which detonation can still take place. The critical diameter is strongly texture-dependent, so that it is larger in cast than in pressed charges. Moreover, finely dispersed gas inclusions considerably reduce the critical diameter. For very insensitive materials such as AN, the critical diameter may be very large.

3.4 Prediction of the detonation velocity of ideal explosives

Amongst the many different detonation parameters, the detonation velocity is one of the salient performance parameters for high explosive molecules and formulations. It is important to have suitable methods for the quick estimation of this parameter, because the detonation velocity helps in selecting, tailoring and understanding the behavior of explosives in terms of blast overpressure, fragmentation, penetration and other expected end-effects. These methods can help chemists to design new explosive molecules – from the myriads of possible combinations of elements – as target futuristic molecules. As is the case for other detonation parameters, many computer codes such as BKW [12], CHEETAH [11], and EXPLO5 [10] have been developed for the calculation of the detonation velocity. Although thermochemical codes are extensively used, they have some limitations such as tedious preparation of the input file, the requirement of a computer, cost of the codes, and other requirements. Fortunately, there are many empirical correlations for predicting the detonation velocity, where ease of hand calculation makes these empirical correlations useful. These empirical relations can provide a suitable pathway for the prediction of the detonation velocity of existing molecules and compositions, as well as for futuristic formulations. They are based on a statistical analysis of the behavior of existing explosives using certain physical and chemical parameters as input. They usually use the molecular formula of the explosive along with the density, heat of formation, and some specific functional groups or molecular moieties. There are several reviews in which these empirical methods have been demonstrated or compared [17, 53–55]. Shekhar [56] compared the predicted detonation velocities of five empirical methods for several conventional and new ideal $C_a H_b N_c O_d$ explosive molecules. Since there are many empirical methods in the literature for the prediction of the detonation velocity, some empirical methods that have wider application and more reliable outputs are introduced here. The available empirical methods can be categorized as a function of different variables:

(i) loading (initial) density, element compositionk and the condensed phase heat of formation of pure or composite explosives;
(ii) loading density, element composition, and the gas phase heat of formation of the pure component;
(iii) loading density and molecular structures of the high explosive;
(iv) maximum attainable detonation velocity.

These methods are reviewed in the following sections.

3.4.1 Detonation velocity as a function of the loading density, element composition, and the condensed phase heat of formation of pure and composite explosives

Most of the available correlations belong to this category, and among the many different methods, two reliable methods are reviewed here.

3.4.1.1 Kamlet and Jacobs (K–J) Method

Kamlet and Jacobs (K–J) [5], as well as Kamlet and Hurwitz [57] used equations (1.5) and (1.10) to obtain the following correlation for the detonation velocity for $C_a H_b N_c O_d$ explosives at loading densities above $1\,\text{g/cm}^3$ as

$$D_{\text{det}} = 3.9712(n'_{\text{gas}})^{0.5}(\bar{M}_{\text{wgas}}Q_{\text{det}}[H_2O(g)])^{0.25}(1 + 1.3\rho_0), \qquad (3.2)$$

where D_{det} is in km/s, n'_{gas} is the number of moles of gaseous detonation products per gram of explosive, $\bar{M}_{\text{wgas}}$ is the average molecular weight of the gaseous products in g/mol, $Q_{\text{det}}[H_2O(g)]$ is in kJ/mol, and ρ_0 is the initial density in g/cm^3. Equation (3.2) confirms that for individual explosives, the experimentally measured values of the detonation velocity are linear with respect to ρ_0. A list of energetic materials containing $C_a H_b N_c O_d$ explosives, as well as their compositions and the corresponding condensed phase heat of formation is given in the Appendix. The detonation velocity of HMX is now calculated as an example. The values of n'_{gas} and $\bar{M}_{\text{wgas}}$ are calculated on the basis of equation (1.10) (a) as

$$C_4 H_8 N_8 O_8 \rightarrow 4N_2 + 4H_2O + 2CO_2 + 2C,$$

$$n'_{\text{gas}} = \frac{[n_{N_2} + n_{H_2O} + n_{CO_2}]}{\text{formula weight of HMX}} = \frac{[4 + 4 + 2]}{296.15}$$

$$= 0.03378,$$

$$\bar{M}_{\text{wgas}} = \frac{[4M_{wN_2} + 4M_{wH_2O} + 2M_{wCO_2}]}{[n_{N_2} + n_{H_2O} + n_{CO_2}]}$$

$$= \frac{[4 \times 28.01 + 4 \times 18.02 + 2 \times 44.01]}{[4 + 4 + 2]}$$

$$= 27.21.$$

The value of $Q_{\text{det}}[H_2O(g)]$ for HMX was calculated in Section 1.2.1, and is equal to $6.18\,\text{kJ/g}$. Since the crystal density of HMX is $1.89\,\text{g/cm}^3$, the use of these values in equation (3.2) provides a calculated detonation velocity of HMX as

$$D_{\text{det}} = 3.9712(n'_{\text{gas}})^{0.5}(\bar{M}_{\text{wgas}}Q_{\text{det}}[H_2O(g)])^{0.25}(1 + 1.3\rho_0)$$

$$= 3.9712(0.03378)^{0.5}(27.12 \times 6.18)^{0.25}(1 + 1.3 \times 1.89)$$

$$= 9.09\,\text{km/s}.$$

The calculated detonation velocity is close to experimental value which is 9.11 km/s [39].

3.4.1.2 Development of the K–J method for $C_aH_bN_cO_dF_eCl_f$ explosives

It was recently shown that the decomposition paths given in equation (1.11) can be used to predict the detonation velocities of $C_aH_bN_cO_dF_eCl_f$ explosives at loading densities above 0.5 g/cm³ since [58]

$$D_{det} = 5.5204(n'_{gas})^{0.5}(\bar{M}_{wgas}Q_{det}[H_2O(g)])^{0.25}\rho_0 + 1.97. \tag{3.3}$$

Equation (3.3) has several advantages over equation (3.2).

(i)   It can be used for those explosives containing fluorine or chlorine atoms in their composition.

(ii) It can be applied to loading densities less than 1 g/cm³.

(iii) It was shown that equation (3.3) gives more reliable predictions for explosives containing fluorine or chlorine atoms. For $C_aH_bN_cO_d$ explosives, the reliability of equations (3.2) and (3.3) is the same.

For HMX at a loading density of 1.89 g/cm³, the values of n'_{gas}, $\bar{M}_{wgas}$, and D_{det} are calculated on the basis of equation (1.11) (c), as well as the calculated value $Q_{det}[H_2O(g)] = 5.02$ kJ/g (Section 1.2.1) as

$$C_4H_8N_8O_8 \rightarrow 4N_2 + 4H_2O + 4CO,$$

$$n'_{gas} = \frac{[n_{N_2} + n_{H_2O} + n_{CO}]}{\text{formula weight of HMX}} = \frac{[4 + 4 + 4]}{296.15}$$

$$= 0.04052,$$

$$\bar{M}_{wgas} = \frac{[4M_{wN_2} + 4M_{wH_2O} + 4M_{wCO}]}{[n_{N_2} + n_{H_2O} + n_{CO}]}$$

$$= \frac{[4 \times 28.01 + 4 \times 18.02 + 4 \times 28.01]}{[4 + 4 + 4]}$$

$$= 24.68,$$

$$D_{det} = 5.5204(n'_{gas})^{0.5}(\bar{M}_{wgas}Q_{det}[H_2O(g)])^{0.25}\rho_0 + 1.97$$

$$= 5.5204(0.04052)^{0.5}(24.68 \times 5.02)^{0.25}(1.89) + 1.97$$

$$= 8.98 \text{ km/s}.$$

The calculated detonation velocity from equation (3.3) is also close to the experimental value which is 9.11 km/s [39].

3.4.2 Detonation velocity as a function of the loading density, element composition, and the gas phase heat of formation of the pure component

A simple correlation has been recently introduced to calculate the detonation velocity of $C_aH_bN_cO_d$ explosives as follows [59]:

$$D_{det} = 1.90 + \left(\frac{-2.97a + 9.32b + 27.68c + 98.9d + 0.292\Delta_fH^\theta(g)}{\text{formula weight of explosive}} \right) \rho_0. \quad (3.4)$$

This correlation has several advantages over equation (3.2).

(i) Equation (3.4) only requires the gas phase heat of formation of the explosive, rather than the condensed phase heat of formation.

(ii) It is a very simple method because it only requires the elemental composition.

(iii) It can be used for loading densities which are less than $1\,\text{g/cm}^3$.

As was seen in Section 1.3.1.1, the values of $\Delta_fH^\theta(g)$ for CL 20 ($C_6H_6N_{12}O_{12}$) and HNS ($C_{14}H_6N_6O_{12}$) are 500 and 150 kJ/mol, respectively. Thus, inserting these values in equation (3.4) gives the following detonation velocities for CL-20 ($\rho_0 = 2.04\,\text{g/cm}^3$) and HNS ($\rho_0 = 1.70\,\text{g/cm}^3$):

$$D_{det} = 1.90 + \left(\frac{-2.97 \times 6 + 9.32 \times 6 + 27.68 \times 12 + 98.9 \times 12 + 0.292 \times 500}{438.18} \right)(2.04)$$

$$= 9.84\,\text{km/s},$$

$$D_{det} = 1.90 + \left(\frac{-2.97 \times 14 + 9.32 \times 6 + 27.68 \times 6 + 98.9 \times 12 + 0.292 \times 150}{450.23} \right)(1.70)$$

$$= 7.23\,\text{km/s}.$$

These calculated values are close to the measured detonation velocities of CL 20 (9.40 km/s [60]) and HNS (7.00 km/s [39]).

3.4.3 Detonation velocity as a function of the loading density and molecular structures of high explosives

It has been shown that the molecular structures of $C_aH_bN_cO_d$ explosives can be used to predict the detonation velocity according to [61]

$$D_{det} = 1.6439 + 3.5933\rho_0$$

$$- 0.1326a - 0.0034b + 0.1206c + 0.0442d$$

$$- 0.2768n_{-NRR'}, \quad (3.5)$$

where $n_{-NRR'}$ is the number of $-NH_2$, NH_4^+, or $\underset{N}{\overset{N}{\sqsubset}}$ groups. This correlation is very simple but it has two limitations:

(i) Deviation from the experimental data increases with increasing addition of nonenergetic additives – as in the case of explosive mixtures.
(ii) This correlation cannot be used for highly or over oxidized explosives such as TNM, as well as mixtures of them with other components, e. g. LX-01.

For PBX-9501 containing 95/2.5/2.5 HMX/Estane/EDNPA-F ($C_{1.47}H_{2.86}N_{2.60}O_{2.69}$) with a loading density of 1.84 g/cm^3, the calculated detonation velocity using equation (3.5) can be obtained as follows:

$$D_{det} = 1.6439 + 3.5933 \times 1.84$$
$$- 0.1326 \times 1.47 - 0.0034 \times 2.86 + 0.1206 \times 2.60 + 0.0442 \times 2.69$$
$$- 0.2768 \times 0$$
$$= 8.48 \text{ km/s.}$$

The estimated detonation velocity is close to the experimental value of 8.83 km/s [62].

3.4.4 Maximum attainable detonation velocity

There are two methods which can be used for the prediction of the detonation velocity at maximum loading – or at the theoretical maximum density. These methods calculate a value for the maximum attainable detonation velocity of pure and composite ideal explosives, and are reviewed here.

3.4.4.1 Rothstein and Peterson's method
Rothstein and Peterson [63, 64] introduced a method suitable for the calculation of the detonation velocity at the theoretical maximum density of the explosive ($D_{det,max}$) for $C_aH_bN_cO_dF_e$ explosives. For ideal explosives, a relationship exists between $D_{det,max}$ with the chemical composition and structure of the high explosive:

$$D_{det,max} = \begin{cases} \dfrac{181.82\left[c + d + e - \frac{b-n(HF)}{2d} + \frac{AB}{3} - \frac{n(B/F)}{1.75} - \frac{n(C=O)}{2.5} - \frac{n(C-O)}{4} - \frac{n(NO_3)}{5}\right]}{\text{formula weight of explosive}} - G \\ \\ - 0.473, \end{cases}$$

(3.6)

where $n(HF)$ is the number of hydrogen fluoride molecules that can possibly be formed from the available hydrogen, $n(B/F)$ is the number of oxygen atoms in excess of those necessary to form CO_2 and H_2O and/or the number of fluorine atoms in excess of those required to form HF, $n(C=O)$ is the number of oxygen atoms doubly bonded directly to carbon, $n(C-O)$ is the number of oxygen atoms singly bonded directly to carbon, $n(NO_3)$ is the number of nitrate groups existing as a nitrate ester, or as a nitric acid salt such as hydrazine mononitrate, G is equal to 0.4 for liquid explosives and 0 for solid explosives, and finally AB is equal to 1 for aromatic compounds and otherwise it is given the value 0. The relation $\frac{[b-n(\text{HF})]}{2d}$ is 0 if $d = 0$, or if $n(\text{HF}) \geq b$. For example, the calculated $D_{\text{det,max}}$ of HMX can be calculated as follows:

$$D_{\text{det,max}} = \left\{ \frac{181.82[8 + 8 + 0 - \frac{8-0}{2\times 8} + \frac{0}{3} - \frac{0}{1.75} - \frac{0}{2.5} - \frac{0}{4} - \frac{0}{5}]}{296.15} - 0 \right\} - 0.473$$

$$= 9.04 \text{ km/s}.$$

The value for the detonation velocity calculated using equation (3.6) is close to experimental value of 9.11 km/s [39].

3.4.4.2 The use of elemental composition and specific structural parameters

It was shown that it is possible to develop a simple and more reliable method for the prediction of $D_{\text{det,max}}$ than that of Rothstein and Peterson's [63, 64]. This methods is as follows [65]:

$$D_{\text{det,max}} = 7.678 - 0.1977a - 0.1105b + 0.2940c + 0.0742d$$
$$- 0.6347n_{\text{NR}} - 0.7354n_{\text{mN}}, \tag{3.7}$$

where n_{NR} is the number of $-N=N-$ groups or NH_4^+ cations in the explosive, and n_{mN} is the number of nitro groups attached to carbon in nitro compounds in which $a = 1$. Equation (3.7) can be easily applied to pure and composite explosives. For example, using equation (3.7) for LX-01, with the composition 51.7/33.2/15.1 NM/TNM/1/Nitro-propane ($C_{1.52}H_{3.73}N_{1.69}O_{3.39}$), gives a value for $D_{\text{det,max}}$ according to

$$D_{\text{det,max}} = 7.678 - 0.1977 \times 1.52 - 0.1105 \times 3.73 + 0.2940 \times 1.69 + 0.0742 \times 3.39$$
$$- 0.6347 \times 0 - 0.7354 \times 4 \times 0.332$$
$$= 6.74 \text{ km/s},$$

which can be compared to the measured value of LX-01 at a maximum loading density of 1.24 g/cm^3 of 6.84 km/s [39].

3.4.5 Comparison of empirical correlations with computer codes

Elbeih et al. [66] measured the detonation velocity of bicyclo-HMX (cis-1,3,4,6-tetra-nitro-octahydroimidazo-[4,5-d] imidazole or BCHMX) as a plastic explosive bonded with the C4 matrix and with Viton A. They have also measured the detonation velocities for a series of nitramines, namely RDX, HMX, and CL 20 with the same types of binders. They have used the CHEETAH [11] and EXPLO5 codes [10] with those equations of state (EOSs) that provide predicted detonation velocities which are in good agreement with those obtained from experimental data. They applied the BKWS-EOS

Table 3.1: Comparison of the outputs of the computer codes CHEETAH and EXPLO5 as well as of several empirical methods with the experimental data for some pure and plastic bonded explosives [66].

Explosive	Formula	$\Delta_f H^\ominus$ (explosive) (kJ/mol)	ρ_0 (g/cm³)	D_{det} (km/s) Exp.	CHEETAH	%Dev	EXPLO5	%Dev
RDX cryst.	$C_3H_6N_6O_6$	47.5	1.76	8.75	8.77	0.2	8.72	-0.3
RDX-C4	$C_{4.66}H_{8.04}N_6O_{5.99}$	22	1.61	8.055	7.96	-1.2	7.86	-2.4
RDX-Viton A	$C_{3.63}H_{6.46}N_6O_{5.95}F_{0.77}$	-90.3	1.76	8.285	8.33	0.5	8.3	0.2
β-HMX cryst	$C_4H_8N_8O_8$	77	1.9	9.1	9.38	3.1	9.23	1.4
HMX-C4	$C_{6.12}H_{11.40}N_8O_{8.14}$	2.6	1.67	8.318	8.27	-0.6	8.13	-2.3
HMX-Viton A	$C_{4.82}H_{8.51}N_8O_{7.93}F_{1.02}$	-75.4	1.84	8.602	8.68	0.9	8.63	0.3
BCHMX−3 % Viton B	$C_{4.23}H_{6.14}N_8O_8F_{0.33}$	173.5	1.79	8.65	8.69	0.5	8.71	0.7
BCHMX-C4	$C_{6.18}H_{10.24}N_8O_{8.08}$	55.3	1.66	8.266	8.15	-1.4	8.06	-2.5
BCHMX-Viton A	$C_{4.85}H_{6.67}N_8O_{7.99}F_{1.02}$	48.6	1.81	8.474	8.47	0.0	8.5	0.3
ε-HNIW	$C_6H_6N_{12}O_{12}$	397.8	1.96	9.44	9.37	-0.7	9.34	-1.1
HNIW-C4	$C_{9.17}H_{12.09}N_{12}O_{12.19}$	201.1	1.77	8.594	8.44	-1.8	8.43	-1.9
HNIW-Viton A	$C_{7.22}H_{6.83}N_{12}O_{11.93}F_{1.51}$	127.7	1.94	9.023	8.77	-2.8	8.85	-1.9

Explosive	D_{det} (km/s) Exp.	K-J	%Dev	Eq. (3.3)	%Dev	Eq. (3.5)	%Dev
RDX cryst.	8.75	8.62	-1.5	8.48	-3.1	8.5	-2.4
RDX-C4	8.055	7.68	-4.7	7.86	-2.4	7.8	-3.5
RDX-Viton A	8.285	—	—	8.46	2.1	—	—
β-HMX cryst	9.1	9.11	0.1	9.02	-0.9	9.2	1.5
HMX − C4	8.318	7.91	-4.9	7.93	-4.7	8.1	-2.4
HMX-Viton A	8.602	—	—	8.85	2.9	—	—
BCHMX−3 % Viton B	8.65	8.64	-0.1	8.67	0.2	8.8	1.9
BCHMX-C4	8.266	7.87	-4.8	7.89	-4.5	8.1	-2.3
BCHMX-Viton A	8.474	—	—	8.64	2.0	—	—
ε-HNIW	9.44	9.37	-0.7	9.29	-1.6	9.8	4.3
HNIW-C4	8.594	8.25	-4.0	8.3	-3.4	8.7	1.6
HNIW-Viton A	9.023	—	—	9.47	5.0	—	—

Table 3.2: The computed outputs using the computer code EXPLO5 and several empirical methods for several new explosives [67].

Explosive	Formula	$\Delta_f H^{\theta}$ (explosive) (kJ/mol)	ρ_0 (g/cm^3)	D_{det} (km/s) EXPLO5	K–J	Eq. (3.3)	Eq. (3.5)
5-Nitrotetrazol-2-ylacetonitrile	$C_3H_2N_6O_2$	495.86	1.747	8.36	7.98	8.15	8.33
5-(5-Nitrotetra-zol-2-ylmethyl) tetrazole monohydrate	$C_3H_5N_9O_3$	297.84	1.796	8.34	8	8.01	8.9
5-(5-Nitrotetra-zol-2-ylmethyl) tetrazole	$C_3H_3N_9O_2$	549.7	1.802	8.68	8.05	8.17	8.88
Ammonium 5-(5-nitrotetra-zol-2-ylmethyl) tetrazolate monohydrate	$C_3H_8N_{10}O_3$	128.15	1.594	7.66	7.12	6.97	8.29
Ammonium 5-(5-nitrotetra-zol-2-ylmethyl) tetrazolate	$C_3H_6N_{10}O_2$	382.9	1.601	7.86	7.2	7.24	8.27
Guanidinium 5-(5-nitrotetra-zol-2-ylmethyl) tetrazolate	$C_4H_8N_{12}O_2$	567.2	1.642	8.06	7.33	7.44	8.52
Amino-guanidinium 5-(5-nitrotetra-zol-2-ylmethyl) tetrazolate	$C_4H_9N_{13}O_2$	615.7	1.633	8.2	7.35	7.45	8.61

and the BKWN-EOS set of parameters for the CHEETAH [11] and EXPLO5 codes [10], respectively. They froze the composition of the detonation products at a temperature of 1800 K on the isentrope beginning at the C–J point. Table 3.1 compares the outputs of the computer codes and several empirical methods with the experimental data for the explosives mentioned above. Table 3.2 also shows the calculated detonation velocities for equations (3.2), (3.3), and (3.5), as well as the computed results obtained using the EXPLO5 code [10] using the BKWN-EOS for several new explosives which have a high nitrogen content, and for which the experimental data of the condensed (solid) phase of formation were available.

3.5 Estimation of the detonation velocity of nonideal explosives

Although aluminum and AN are widely used in military and commercial explosives, assessment of their detonation performance is difficult. In computer codes, the amount of initial AN and aluminum that is assumed to react can be specified by invoking their partial equilibrium [39]. For example, inert Al atoms that can only form solid, liquid, or gaseous Al can be included in the product species data base [39]. This situation prevents the reaction of aluminum with oxygen or other reactive species. Increasing the number of gaseous products and decreasing the amount of condensed carbon can be improved by prevention of the formation of Al_2O_3. Thus, the detonation velocity increases with increasing gas yield. If complete equilibrium is assumed, a higher amount of condensed Al_2O_3 is produced, because this situation forces oxygen to react with aluminum rather than carbon. On the oxidation of aluminum, increased temperature produces a hot, fuel-rich gaseous phase and more solid carbon, because the high temperature is a result of the large negative heat of formation of Al_2O_3. Besides computer codes, there are several empirical methods for the prediction of the detonation velocity of nonideal aluminized and nitrated salt explosives. Among these methods, several methods that have wider applications in different formulations are reviewed here.

3.5.1 Detonation velocity of ideal and nonideal explosives as a function of the loading density, element composition, and the condensed phase heat of formation of pure or composite explosives

Since the detonation velocity of an energetic compound depends on n'_{gas}, $\bar{M}_{wgas}$, and $Q_{det}[H_2O(g)]$, the decomposition paths of equation (1.23) can control these parameters to obtain a reliable correlation for the calculation of the detonation velocity of ideal and nonideal explosives.

Thus, it was shown that the following equation can be used to predict the detonation velocity of ideal and non-ideal explosives with the general formula $C_aH_bN_cO_d F_eCl_fAl_g(NH_4NO_3)_h$ [27]:

$$D_{det} = 5.468(n'_{gas})^{0.5}(\bar{M}_{wgas}Q_{det}[H_2O(g)])^{0.25}\rho_0 + 2.05. \qquad (3.8)$$

Table 3.3 shows the calculated data for several pure and composite explosives. Equation (3.8) is an improved method of that given in Section 3.4.1.2, which can be applied to ideal $C_aH_bN_cO_dF_eCl_f$ explosives as well as nonideal aluminized and AN explosives.

Table 3.3: Several examples of the calculation of the detonation velocities for different ideal or nonideal high explosives.

Path	Name	ρ_0 (g/cm³)	$D_{Exp.}$ (km/s)	moles of products/moles of high explosive											
				HF	HCl	N_2	CO	CO_2	H_2	H_2O	O_2	C(s)	Al_2O_3(s)	Al(s)	AN(s)
2a	AN (Pure)	1.05	4.50	–	–	0.78	–	–	–	1.56	0.39	–	–	–	0.22
2b	AN/Al (70/30)	1.05	5.40	–	–	0.81	–	–	1.62	0.01	0.41	–	0.54	0.03	0.06
2c	LX-17	1.91	7.63	0.2	0.05	1.08	2.15	–	0.96	–	–	0.14	–	–	–
2d	DEGN	1.38	6.76	–	–	1.00	4.00	–	1.00	3.00	–	–	–	–	–
2e	ANFO 6/94	0.88	5.50	–	–	1.00	0.09	0.28	–	2.36	–	–	–	–	–
2f	NG	1.59	7.58	–	–	1.50	–	3.00	–	2.50	0.25	–	–	–	–

Path	Name	n'_{gas} (mol/g)	$\bar{M}_{wgas}$ (g/mol)	$Q_{det}[H_2O(g)]$ (kJ/g)	D_{det} (km/s)
2a	AN (Pure)	0.0341	22.87	1.12	4.43
2b	AN/Al (70/30)	0.0285	13.80	6.17	4.99
2c	LX-17	0.0444	22.12	1.95	7.69
2d	DEGN	0.0459	21.79	3.83	6.93
2e	ANFO 6/94	0.0437	23.03	3.69	5.10
2f	NG	0.0319	32.15	6.23	7.89

3.5.2 Using molecular structure to predict the detonation velocity of ideal and nonideal explosives

It was found that detonation velocity of explosives containing aluminum and nitrated salts can be given as [31]

$$D_{det} = 1.64 + 3.65\rho_0 - 0.135a + 0.117c + 0.0391d$$
$$- 0.295n_{NR_1R_2} - 0.620n'_{Al} - 1.41n'_{NO_3 \text{ salt}}, \tag{3.9}$$

where $n_{NR_1R_2}$ is the number of specific groups in explosives, and n'_{Al} and $n'_{NO_3 \text{ salt}}$ are two correcting functions which show the contribution of the number of moles of Al and the number of moles of nitrate salt in composite explosives under certain conditions, respectively. The specific group NR_1R_2 includes $-NH_2$, NH_4^+ and five membered rings with three (or four) nitrogens in any explosive, as well as five (or six) membered rings in nitramine cages. The value of n'_{Al} is equal to the number of moles of aluminum except its value should be corrected for the following conditions:
(i) if $d \leq a + 0.1$, then $n'_{Al} = 0.75n_{Al} + 1.00$,
(ii) if $d \geq a + \frac{b}{2}$, then $n'_{Al} = n_{Al} - 0.375$;

and $n'_{NO_3 \text{ salt}}$ is equal to the number of moles of nitrate salt with the following exceptions:
(iii) if $d \leq a + \frac{3b}{5}$ then $n'_{NO_3 \text{ salt}} = n_{NO_3 \text{ salt}} - 1.50$,
(iv) If $d \geq 2a + \frac{b}{4}$ then $n'_{NO_3 \text{ salt}} = 1.75n_{NO_3 \text{ salt}}$.

To use equation (3.9) for aluminized explosives, 100 g of the explosive was used for calculation of the detonation velocity, and the number of moles of extra aluminum in the explosives with general formula $C_aH_bN_cO_dAl_e$ should be considered, e. g. TNT/Al (89.4/10.6) has the formula $C_{2.756}H_{1.968}N_{1.181}O_{2.362}Al_{0.3929}$, which, on the basis of a 100 g mixture of TNT and Al should be changed to the formula $C_{3.084}H_{2.203}N_{1.322}O_{2.643}Al_{0.440}$, i. e.

$$\text{C:} \quad 2.756 \times \frac{100}{89.4} = 3.0841;$$

$$\text{H:} \quad 1.968 \times \frac{100}{89.4} = 2.203;$$

$$\text{N:} \quad 1.181 \times \frac{100}{89.4} = 1.322;$$

$$\text{O:} \quad 2.362 \times \frac{100}{89.4} = 2.643;$$

$$\text{Al:} \quad 0.3929 \times \frac{100}{89.4} = 0.440.$$

Thus, the detonation velocity of TNT/Al (89.4/10.6) at a loading density of 1.72 g/cm³ is calculated on the basis of condition (i) as

$$D_{det} = 1.64 + 3.65 \times 1.72 - 0.135 \times 3.084 + 0.117 \times 2.203$$
$$+ 0.0391 \times 2.643 - 0.295 \times 0 - 0.620[0.75 \times 0.44 + 1.00] - 1.41 \times 0$$
$$= 6.94 \text{ km/s}.$$

The predicted value is close to the measured value which is 7.05 km/s [39].
Equation (3.9) has two limitations:
(i) this new procedure cannot be used for highly over-oxidized explosives, e. g. TNM, or for their mixtures with other components such as LX-01;
(ii) deviation from the experimental data increases with increasing use of nonenergetic additives.

3.5.3 Maximum attainable detonation velocity of $C_aH_bN_cO_dF_e$ and aluminized explosives

For $C_aH_bN_cO_dF_e$ and aluminized explosives, it was shown that equation (3.10) can be used to predict $D_{det,max}$ as [32]:

$$D_{det,max} = 7.03 - 0.162a - 0.0206b + 0.228c + 0.0714d$$
$$+ 0.996D_{det,max}^{Inc} - 0.741D_{det,max}^{Dec}, \tag{3.10}$$

where $D_{det,max}$ is expressed in km/s; $D_{det,max}^{Inc}$ and $D_{det,max}^{Dec}$ are two correcting functions that increase and decrease the predicted results on the basis of a, b, c, and d by taking into account the content of fluorine and Al. For pure and composite explosives, the values of $D_{det,max}^{Inc}$ and $D_{det,max}^{Dec}$ are given in the following sections.

3.5.3.1 Pure explosives
(a) Prediction of $D_{det,max}^{Inc}$
(i)   The value of $D_{det,max}^{Inc}$ is equal to 0.8 in presence of one of the following groups: $-NH-NO_2$, more than one $-NNO_2$, nitrate derivative of hydrazine or $N-C(NO_2)-N$ in cyclic heteroarenes.
(ii) For fluorinated aromatic compounds, $D_{det,max}^{Inc}$ is equal to 0.5.
(iii) For explosives with $b = 0$, the value of $D_{det,max}^{Inc}$ equals 1.0.
(b) Prediction of $D_{det,max}^{Dec}$
(i)   In the presence of one $-C(=O)-C(=O)-$ and two $-C(=O)-$ groups, $D_{det,max}^{Dec}$ equals 1.75, and with one ether group, the value of $D_{det,max}^{Dec}$ is 0.7.
(ii) For explosives with $a = 1$, the value of $D_{det,max}^{Dec}$ is equal to the number of nitro groups attached to carbon atom (n_{NO_2}).
(iii) For CHNF explosives, the value of $D_{det,max}^{Dec}$ is equal to 1.
(iv) The value of $D_{det,max}^{Dec}$ is 1.7 if the 1,3,5-triazine or pyrimidine rings are present in polynitro heteroarenes.

3.5.3.2 Composite explosives

(a) Ideal explosives
  (i) Nitramine composite explosives: For nitramines with one $-NH-NO_2$ or more than one $-NNO_2$ as well as a nitrate derivative of hydrazine, the values of $D_{det,max}^{Inc}$ depend on the weight percent of high explosives (WPHE):
  (a) if WPHE $\geq$ 85, then $D_{det,max}^{Inc} = 1.1$;
  (b) if $50 \leq$ WPHE < 85, then $D_{det,max}^{Inc} = 0.8$;
  (c) if WPHE < 50, then $D_{det,max}^{Inc} = 0.0$.
  (ii) With TNM and NM: $D_{det,max}^{Dec} = 1.0$; for other liquid mixtures of TNM: $D_{det,max}^{Dec} = 0.5$.

(b) Aluminized explosives
  (i) For nitramines with more than one $-NNO_2$, the ratio of the number of moles of explosive (n_{exp}) to Al (n_{Al}) can be used to find $D_{det,max}^{Inc}$ and $D_{det,max}^{Dec}$:
  (a) if $n_{exp}/n_{Al} > 0.13$, then $D_{det,max}^{Inc} = n_{exp}/n_{Al}$;
  (b) if $n_{exp}/n_{Al} \leq 0.13$, then $D_{det,max}^{Dec} = 0.4$.
  (ii) For high explosives with a high enough oxygen content, i. e. $d > a + b/2$, the value of $D_{det,max}^{Inc}$ is 0.70.

Table 3.4 shows a summary of the above conditions. Equation (3.10) can be used for pure and composite $C_aH_bN_cO_dF_e$ explosives as well as for aluminized explosives. Thus, the wider application of equation (3.10) with respect to equations (3.6) and (3.7) is the main advantage of the present method. For example, the value of $D_{det,max}$ of 74.766/18.691/4.672/1.869 TNT/Al/Wax/Graphite with the general formula $C_{2.79}H_{2.31}N_{0.99}O_{1.98}Al_{0.69}$ at a loading density of 1.68 g/cm^3 is given as follows:

$$D_{det,max} = 7.03 - 0.162 \times 2.79 - 0.0206 \times 2.31 + 0.228 \times 0.99 + 0.0714 \times 1.98$$
$$+ 0.996 \times 0 - 0.741 \times 0$$
$$= 6.90 \text{ km/s.}$$

The predicted detonation velocity is close to the measured value of 6.50 km/s [12].

3.6 Assessment of the detonation velocity of primary explosives

Primary explosives are hazardous energetic materials with high sensitivities to different stimuli such as impact, friction, flame, heat, electric spark, and shock. Since primary explosives generate a large amount of heat and/or shockwave, they are usually used to initiate the detonation of high explosives or to ignite propellants and pyrotechnics [68]. The detonation velocity of a primary explosive shows a significant effect on its ignition energy and spark sensitivity [69]. For a homogeneous military-type explosive in large diameter cylinders, the detonation velocity is a linear function of the initial density of the explosive in the range of $\rho > 0.80$ g/cc [70]. Since common pri-

Table 3.4: Summary of the estimated values of $D_{det,max}^{Inc}$ and $D_{det,max}^{Dec}$.

Explosive	Functional groups or structural moieties	$D_{det,max}^{Inc}$	$D_{det,max}^{Dec}$	Condition
Pure $C_aH_bN_cO_dF_e$ explosives				
Nitramine	one $-NH-NO_2$ or more than one $-NNO_2$	0.8	0.0	—
Nitrate derivative of hydrazine	—			
Cyclic heteroarenes	$N-C(NO_2)-N$			
Fluorinated aromatic compounds	fluorine attached to aromatic ring	0.5	0.0	—
Energetic compounds without hydrogen atom	—	1.0		—
Energetic compounds containing carbonyl or one etheric groups	one $-C(=O)-C(=O)-$	0.0	1.75	—
	two $-C(=O)-$ or one etheric groups	0.0	0.7	—
Nitro compounds with one carbon	—	0.0	n_{NO_2}	—
CHNF explosives	—	0.0	1.0	—
Polynitro heteroarenes	1,3,5-triazine or pyrimidine ring	0.0	1.7	—
Mixture of high explosives on the basis of 100 g				
Solid mixture				
Nitramine or nitrate derivative of hydrazine	one $-NH-NO_2$ or more than one $-NNO_2$	1.1	0.0	WPHE $\geq$ 85
		0.8	0.0	50 $\leq$ WPHE < 85
		0.0	0.0	WPHE < 50
Aluminized explosives	more than one $-NNO_2$	n_{exp}/n_{Al}	0.0	n_{exp}/n_{Al} > 0.13
		0.0	0.4	n_{exp}/n_{Al} $\leq$ 0.13
	—	0.7	0.0	for explosives with $d > a + b/2$
Liquid mixture				
Mixture TNM and NM	—	1.0	0.0	—
Mixture TNM and the other organic compound	—	0.5	0.0	—

mary explosives contain C, H, N, O, Ni, Pb, Ag, and Hg atoms, decomposition products of various primary explosives with the general formula of $C_aH_bN_cO_dNi_ePb_fAg_gHg_h$ were used to derive a suitable correlation for the prediction of detonation velocity of primary explosives [71]. Equation (3.11) assumes four reaction paths in which metal atoms appear in their standard state, all nitrogen atoms convert to N_2, carbon atoms in oxygen-lean compounds go to graphite and CO while in oxygen-rich compounds they go to CO and CO_2, and finally, hydrogen and oxygen atoms participate in formation of H_2, H_2O and O_2 products [71].

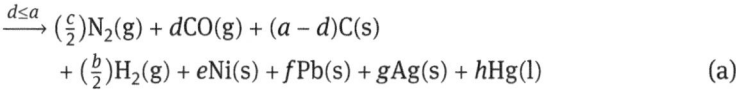

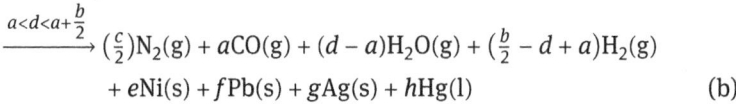

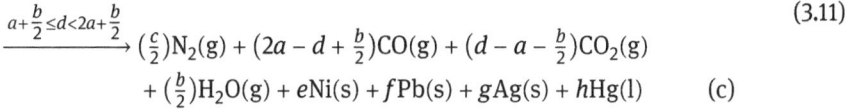

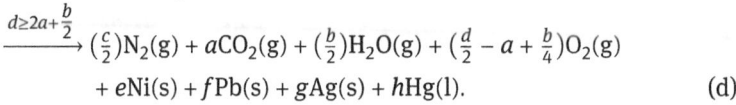

$$\xrightarrow{d \le a} (\tfrac{c}{2})N_2(g) + dCO(g) + (a-d)C(s)$$
$$+ (\tfrac{b}{2})H_2(g) + eNi(s) + fPb(s) + gAg(s) + hHg(l) \tag{a}$$

$$\xrightarrow{a < d < a + \tfrac{b}{2}} (\tfrac{c}{2})N_2(g) + aCO(g) + (d-a)H_2O(g) + (\tfrac{b}{2} - d + a)H_2(g)$$
$$+ eNi(s) + fPb(s) + gAg(s) + hHg(l) \tag{b}$$

$$\tag{3.11}$$

$$\xrightarrow{a + \tfrac{b}{2} \le d < 2a + \tfrac{b}{2}} (\tfrac{c}{2})N_2(g) + (2a - d + \tfrac{b}{2})CO(g) + (d - a - \tfrac{b}{2})CO_2(g)$$
$$+ (\tfrac{b}{2})H_2O(g) + eNi(s) + fPb(s) + gAg(s) + hHg(l) \tag{c}$$

$$\xrightarrow{d \ge 2a + \tfrac{b}{2}} (\tfrac{c}{2})N_2(g) + aCO_2(g) + (\tfrac{b}{2})H_2O(g) + (\tfrac{d}{2} - a + \tfrac{b}{4})O_2(g)$$
$$+ eNi(s) + fPb(s) + gAg(s) + hHg(l). \tag{d}$$

Table 3.5 shows the predicted values of n'_{gas}, $\bar{M}_{wgas}$ and $Q_{det}[H_2O(g)]$ on the basis of equation (3.11) for various primary explosives.

Equation (3.12) shows a suitable correlation for estimation of the detonation velocity of primary explosives [11].

$$D_{det} = 2.081 + 0.1144\rho_0 + 8.055(n'_{gas})(\bar{M}_{wgas}Q_{det}[H_2O(g)])^{0.5}\rho_0 \tag{3.12}$$

For example, equation (3.12) gives the value of 5.96 km/s for 1-(5-Tetrazolyl)-3-guanyltetrazene hydrate, or tetrazene, using data given in Table 3.5 at $\rho_0 = 1.63\,g/cm^3$. The calculated detonation velocity by the Explo5 V6.02 thermochemical computer code [10, 77] using the BKW G-S equation of states gives the value of 6.8 km/s [77]. The measured value of the detonation velocity for tetrazene is 5.30 km/s [78]. It was shown that the **mean absolute percentage error** (MAPE) of the new model for primary explosives given in Table 3.5 (corresponding to 72 data points) is 8.73, which is lower than MAPE value of Explo5 code's predictions, i. e. 13.43 [11].

A hybridization of support vector regression (SVR) and genetic algorithm (GA) have also been used for estimation of the detonation velocities of primary explosives [79] based on experimental data of Jafari et al. [71]. This approach requires complex computer codes and expert users.

Summary

Different methods for the prediction of the detonation velocities of ideal and nonideal explosives have been reviewed in this chapter. The best available predictive methods have been introduced. The main difficulties lie in the uncertainty of the degree of aluminum and AN oxidized at the C–J point for mixture of high explosive with aluminum and AN. For oxygen-poor explosives, it can be assumed that a small amount of aluminum can react with detonation products. Furthermore, a higher percentage of AN

Table 3.5: Chemical formula, enthalpies of formation, and properties of detonation products according to equation (3.11).

Composition	Chemical formula	$\Delta_f H^\theta (c)^a$ (kJ/mol)	$n'_{gas} \times 10^3$ (mol/g)	$\bar{M}_{wgas}$ (mol/g)	$Q_{det} [H_2O(g)]$ (kJ/g)
Silver azide	AgN_3	308.8	10.0	28.02	2.06
Silver acetylide – Silver nitrate	$Ag_2C_2 \cdot AgNO_3$	270.0	0.61	35.21	1.89
Nickel hydrazine nitrate	$Ni[(N_2H_4)_3](NO_3)_2$	−448.7	35.9	22.02	3.59
Lead azide	$Pb(N_3)_2$	468.6	10.3	28.02	1.61
Mercury fulminate	$C_2N_2O_2Hg$	268.0	10.5	28.01	1.72
1-(5-Tetrazolyl)-3-guanyltetrazene hydrate	$C_2H_8N_{10}O$	189.1	53.1	17.62	1.59
2,4,6-Triazido-1,3,5-triazine	C_3N_{12}	1 053.0	29.4	28.02	5.16
5,5'-Diazido-1H,1'H-3,3'-bi(1,2,4-triazole)	$C_4H_2N_{12}$	971.0	32.1	24.31	4.45
2-Diazo-4,6-dinitrophenol	$C_6H_2N_4O_5$	321.0	38.1	24.76	4.16
Lead trinitroresorcinate monohydrate	$C_6H_3N_3O_9Pb$	−836.9	19.2	29.35	1.31
3,3,6,6-Tetramethyl-1,2,4,5-tetraoxane	$C_6N_{12}O_4$	−431.6	32.9	28.02	0.03
1,3,5-Triazido-2,4,6-trinitrobenzene	$C_6N_{12}O_6$	1 130.0	35.7	28.02	5.33
1,6-Diaza-3,4,8,9,12,13-hexaoxa-bicyclo[4,4,4]tetradecane	$C_6H_{12}N_2O_6$	−335.0	62.4	16.02	1.58
3,3,6,6,9,9-Hexamethyl-1,2,4,5,7,8-hexaoxonane	$C_9H_{18}O_6$	151.4	57.1	8.52	0.72

[a] Experimental enthalpies of formation from Refs. [1, 72–76].

in mixtures with oxygen-poor explosives may decrease the detonation velocity. The methods introduced in this chapter can be used to estimate the detonation velocities of ideal as well as of nonideal explosives to within about a few percent of the experimental values from the chemical formula of a real or hypothetical mixture.

Primary explosives are hazardous energetic materials with considerable sensitivities to external stimuli. Equation (3.12) has been introduced for prediction of the detonation velocities of different hazardous primary explosives.

Questions and problems

The necessary information for some problems are given in the Appendix. [i]

(1) If the measured detonation velocity of LX-10 at loading density $1.86\,g/cm^3$ is $8.82\,km/s$,
    (a) calculate its detonation velocity using equations (3.2) and (3.3);
    (b) compare the percent deviation of the calculated data.
(2) If the gas phase heat of formation of Oclotol-76/23 is $11.15\,kJ/mol$, use equation (3.4) to calculate its detonation velocity at loading density $1.81\,g/cm^3$.
(3) Use equation (3.5) to calculate the detonation velocity of TNTAB at loading density $1.74\,g/cm^3$.
(4) Use equations (3.6) and (3.7) to calculate the maximum attainable detonation velocity of picric acid which has the following molecular structure:

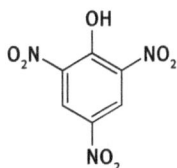

(5) Use equation (3.8) to calculate the detonation velocity of TNTEB/Al(90/10) at loading density $1.75\,g/cm^3$.
(6) Use equation (3.9) to calculate the detonation velocity of AMATEX-20 at loading density $1.66\,g/cm^3$.
(7) Calculate the maximum attainable detonation velocity of PBXC-117 using equation (3.10).

For answers and solutions, please see p. 120 [i]

4 Detonation pressure

Detonation pressure is one of the most important detonation parameters because for many years it has been regarded as one of the principal measures of the performance of a detonating explosive. It is important to predict the time-independent state of chemical equilibrium, which is defined in accordance with the C–J condition. It is reasonable to expect the calculated and the experimental C–J pressures to differ by 10 to 20 % because of the nonsteady-state nature of the detonation wave [12]. For non-ideal explosives, detonation pressures are significantly different from those predicted by equilibrium, one dimensional and steady state calculations.

4.1 Relationship between the detonation pressure and the detonation velocity

In the detonation process, a supersonic propagation of the chemical reaction through an explosive takes place as is shown in Figure 4.1, According to the **Zeldovich–von Neumann-Doering** (**ZND**) model of detonation, chemical reactions occur under the action of a shock wave at a definite rate in the chemical reaction zone [80]. Figure 4.2 shows the detonation wave structure on the basis of the ZND model of detonation, which includes

(a) the shock front followed by the chemical reaction zone, the so-called **chemical spike** or **von Neumann spike;**
(b) the steady chemical reaction zone;
(c) the C–J point;
(d) the **Taylor wave** of isentropic expansion of the detonation products.

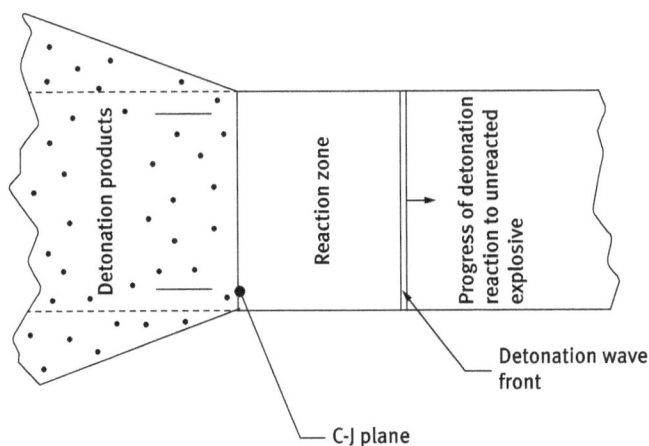

Figure 4.1: Supersonic propagation of a chemical reaction through an explosive.

https://doi.org/10.1515/9783110677652-004

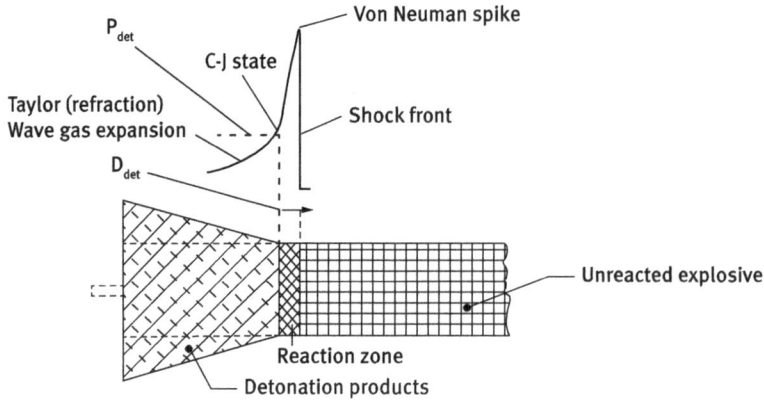

Figure 4.2: The structure of the detonation wave and the propagation of the detonation reaction.

Due to the extreme temperature and pressure conditions, the chemical reaction of an explosive should occur immediately at the wave front. The energy of the reaction can maintain propagation of the shock wave. Thus, the detonation pressure can be given at the C–J point, on the basis of the **momentum balance,** as follows [80]:

$$P_{\text{det}} = \rho_0 D_{\text{det}} W_{\text{C-J}}, \tag{4.1}$$

where P_{det} is the detonation pressure, and $W_{\text{C-J}}$ is the velocity of gaseous products (fumes) at the C–J point. It should be mentioned that the pressure of the original explosive (normally ambient) is small compared to P_{det}, which is neglected in equation (4.1). Thus, the detonation pressure increases very considerably if the initial density of the explosive can be raised to its maximum value. Conversely, the detonation pressure and the detonation velocity may be reduced by employing a more loosely textured explosive. According to Figure 3.1, the **adiabatic exponent** (γ) is defined as the negative of the logarithmic slope of the adiabat [12]:

$$\gamma = -\left(\frac{\partial \ln P}{\partial \ln \frac{1}{\rho}}\right)_S. \tag{4.2}$$

Thus, the adiabatic exponent can be assessed by the initial pressure-volume slope in the isentropic gases from the C–J point, which is primarily a function of ρ_0. The parameter y can also be given at the C–J point as follows [80]:

$$y = \frac{1}{\frac{\rho_0}{\rho_{\text{C-J}}} - 1}, \tag{4.3}$$

where $\rho_{\text{C-J}}$ is the density at C–J point. Combining equations (4.3) and (4.1) with the following mass balance equation of the **Rankine–Huguniot Jump,** i. e. [80]

$$\frac{\rho_0}{\rho_{\text{C-J}}} = \frac{D_{\text{det}} - W_{\text{C-J}}}{D_{\text{det}}},$$

gives the following equation [12]:

$$P_{det} = \frac{\rho_0 D_{det}^2}{\gamma + 1}. \tag{4.4}$$

As is seen in equation (4.4), knowledge of γ can help to estimate the value of P_{det} from D_{det}^2, or vice versa, at the specified loading density. The parameter γ is relatively insensitive to elemental composition and is widely used in the field of explosives. Several suitable correlations have been developed to predict γ as a function of the loading density [81–84] in which it is assumed that γ is independent of the chemical composition. Kamlet and Short [83] introduced a suitable correlation for the calculation of γ in the form of "rule for gamma". Their correlation can be used as a criterion in choosing among conflicting experimental measurements of detonation properties of a number of $C_a H_b N_c O_d$ explosives at a loading density which is greater than $1\,g/cm^3$. Among the available correlations, the following correlation [61] shows good agreement with respect to the corresponding measured values from reported empirical methods [81–84]:

$$\gamma = 1.819 - \frac{0.196}{\rho_0} + 0.712\rho_0. \tag{4.5}$$

Since data for measured detonation velocities is generally more widely available than the detonation pressures for ideal explosives, it is possible to use equations (4.4) and (4.5) to obtain reliable estimations of detonation pressures over a wide range of loading densities, e. g. $0.2–2\,g/cm^3$. For example, EDC-24, 95/5 HMX/Wax with a loading density of $1.705\,g/cm^3$ has a measured detonation velocity of $8300\,m/s$. Incorporation of these values to equations (4.4) and (4.5) give γ and P_{det} as

$$\gamma = 1.819 - \frac{0.196}{1.776} + 0.712(1.776)$$

$$= 2.973,$$

$$P_{det} = \frac{(1.776\frac{g}{cm^3} \times \frac{1\,kg}{1000\,g} \times \frac{10^6\,cm^3}{1\,m^3})(8713\frac{m}{s})^2}{2.973 + 1}$$

$$= 3.394 \times 10^{10}\ Pa$$

$$= 339.4\ kbar.$$

The predicted detonation pressure is close to the measured detonation pressure, i. e. 334 kbar [12].

4.2 Measurement of the detonation pressure

In contrast to the detonation velocity, which can typically be measured to within a few percent, measurements of the detonation pressure and temperature are less accurate. Due to the existence of nonequilibrium effects in reaction zones, the detonation pressures which are determined span a range of 10–20 % [12]. For example, the maximum percent of deviation between the lowest to highest values from the interpretation of different types of measurements for COMP B at 1.73 g/cc in the same laboratory is 16.4 % [85]. The measurement of the pressure at the C–J point and the duration of the chemical reactions in the reaction zone are usually performed through various dynamic methods based on different physical principles. The experimental methods for the measurement of the detonation wave parameters can be classified into two categories:

(i) *Internal methods.* Detonation parameters are directly determined in these methods. For example, the direct determination of the detonation pressure can be achieved using a manganin pressure gauge [52]. Although the time resolution is approximately on the nanosecond scale, it is not sufficient for a reliable study of the narrow reaction zone.

(ii) *Those methods that are based on the registration of the state originating after the shock wave reflection from a barrier.* For example, the determination of the detonation pressure can be achieved using optical methods or electrocontact probes and the oscilloscope technique. Details of these categories are discussed elsewhere [52].

4.3 Estimation of the detonation pressure of ideal explosives

Besides equation (4.4) and the computer codes that were described in previous chapters, there are some empirical methods which can be used to predict the detonation pressure of ideal explosives and nonideal aluminized explosives. The empirical methods which are available can be categorized as a function of different variables, as is shown in the following sections (as was also the case for the detonation velocity).

4.3.1 Detonation pressure as a function of the loading density, element composition, and the condensed phase heat of formation of pure or composite explosives

There are various approaches reported in the open literature for the prediction of the detonation pressure as a function of different variables, including the loading density, elemental composition, and the condensed heat of formation of pure or composite explosives. Among these methods, several which have wide application are reviewed here.

4.3.1.1 The use of the Kamlet and Jacobs (K–J) method
Kamlet and Jacobs (K–J) [5], as well as Kamlet and Ablard [86] used equations (1.5) and (1.10) to obtain a relationship for the detonation pressure for $C_aH_bN_cO_d$ explosives

at loading densities above $1\,g/cm^3$ as

$$P_{det} = 240.86(n'_{gas})(\bar{M}_{wgas}Q_{det}[H_2O(g)])^{0.5}\rho_0^2, \qquad (4.6)$$

where P_{det} is in kbar. Kamlet and Dickinson [87] showed that equation (4.6) could be established using trial-and-error fitting of highly accurately computed detonation pressures. Kazandjian and Danel [88] have indicated that, with the assumptions considered by K–J method [5], the detonation pressure is mainly proportional to

$$(n'_{gas})(\bar{M}_{wgas}Q_{det}[H_2O(g)])^{0.5}\rho_0^2.$$

Equation (4.6) confirms that for individual explosives the measured detonation pressure values are proportional to ρ_0^2. As an example, the detonation pressure of HMX at $1.89\,g/cm^3$ will be calculated for which the values of n'_{gas} and $\bar{M}_{wgas}$ were calculated in Section 3.4.1.1, and the value of $Q_{det}[H_2O(g)]$ for HMX was calculated in Section 1.2.1, and is equal to 6.18 kJ/g:

$$
\begin{aligned}
P_{det} &= 240.86(n'_{gas})(\bar{M}_{wgas}Q_{det}[H_2O(g)])^{0.5}\rho_0^2 \\
&= 3.9712(0.03378)(27.12 \times 6.18)^{0.5} \times 1.89^2 \\
&= 376.9\,kbar.
\end{aligned}
$$

The measured detonation pressure is 390 kbar [39]. Thus, the percent of deviation of the calculated and the measured detonation pressures is −3.5 %.

4.3.1.2 The modified K–J method for $C_aH_bN_cO_dF_eCl_f$ explosives

As was the case for the detonation velocity, the decomposition paths given in equation (1.11) were used to predict the detonation pressures of $C_aH_bN_cO_dF_eCl_f$ explosives at loading densities above $0.8\,g/cm^3$ as [89]:

$$P_{det} = 245.5(n'_{gas})(\bar{M}_{wgas}Q_{det}[H_2O(g)])^{0.5}\rho_0^2 - 11.2. \qquad (4.7)$$

The advantages which were mentioned in Section 3.4.1.2 can also be applied to equation (4.7).

For HMX at loading density $1.89\,g/cm^3$ and using the calculated values of n'_{gas}, $\bar{M}_{wgas}$, and $Q_{det}[H_2O(g)]$ for HMX given in Section 3.4.1.2, the detonation pressure of HMX can be calculated according to:

$$
\begin{aligned}
P_{det} &= 245.5(n'_{gas})(\bar{M}_{wgas}Q_{det}[H_2O(g)])^{0.5}\rho_0^2 - 11.2 \\
&= 245.5(0.04052)(24.68 \times 5.02)^{0.5}(1.89)^2 - 11.2 \\
&= 384.3\,kbar.
\end{aligned}
$$

The percent deviation of the calculated detonation pressure from the measured value is −1.5 % [39].

4.3.2 Detonation pressure as a function of the loading density, element composition and gas phase heat of formation of the pure component

It is possible to predict the detonation pressure of $C_aH_bN_cO_d$ explosives, in the same way as for the detonation velocity given in Section 3.4.2, as follows [90]:

$$P_{det} = -2.6 + \left(\frac{1026a + 226b + 1031c + 3150d + 30.7\Delta_f H^\theta(g)}{M_w} \right) \rho_0^2. \tag{4.8}$$

Three advantages which were mentioned in Section 3.4.2 can also be applied to equation (4.8). It is important to use reliable methods for the estimation of $\Delta_f H^\theta(g)$. For example, 2,4,6-triamino-1,3,5-trinitrobenzene (TATB) with the empirical formula $C_6H_6N_6O_6$ is a well-known, thermally stable explosive. The $\Delta_f H^\theta(g)$ of TATB has been calculated using the B3LYP/6-31G* and semi-empirical method of PM3 which are 30.08 and -45.18 kJ/mol, respectively [91]. Thus, the use of these data for TATB ($\rho_0 = 1.86$ g/cm^3) give the following the following detonation pressures:

$$P_{det} = -2.6 + \left(\frac{1026 \times 6 + 226 \times 6 + 1031 \times 6 + 3150 \times 6 + 30.7 \times 30.08}{258.15} \right)(1.86)^2$$

$$= 281.8 \text{ kbar},$$

$$P_{det} = -2.6 + \left(\frac{1026 \times 6 + 226 \times 6 + 1031 \times 6 + 3150 \times 6 + 30.7(-45.18)}{258.15} \right)(1.86)^2$$

$$= 250.8 \text{ kbar}.$$

Since B3LYP/6-31G* gives more reliable $\Delta_f H^\theta(g)$ values than PM3, the predicted detonation pressure using B3LYP/6-31G* is unsurprisingly closer to the experimental value of 291 kbar [39].

4.3.3 Detonation pressure as a function of the loading density and molecular structure of high explosives

It was shown that it is possible to predict the detonation pressure of $C_aH_bN_cO_d$ explosives using [92]

$$\begin{aligned} P_{det} = {} & -22.32 + 104.04\rho_0^2 \\ & - 10.981a - 1.997b + 5.562c + 5.539d \\ & - 23.68n_{-NH_x} - 154.1n_1^0, \end{aligned} \tag{4.9}$$

where n_{-NH_x} is the number of $-NH_2$ and NH_4^+ in the energetic compounds and n_1^0 equals 1.0 for energetic compounds that follow the condition $d > 3(a + b)$. Equation (4.9) follows two limitations of equation (3.5). As is indicated in equation (4.9), there is no need to use the condensed or gas phase heat of formation value of the explosive. For example, applying equation (4.9) for Octol-60/40, with composition 60/40 HMX/TNT ($C_{2.04}H_{2.50}N_{2.15}O_{2.68}$) at a loading density of 1.80 g/cm^3 gives the detonation pressure as

$$
\begin{aligned}
P_{det} = {} & -22.32 + 104.04 \times 1.80^2 \\
& -10.981 \times 2.04 - 1.997 \times 2.50 + 5.562 \times 2.15 + 5.539 \times 2.68 \\
& -23.68 \times 0 - 154.1 \times 0 \\
= {} & 309\ \text{kbar.}
\end{aligned}
$$

The estimated detonation pressure is close to the value of 320 kbar which was obtained from experiments [39].

4.3.4 Maximum attainable detonation pressure

It was shown that the detonation pressure at the maximum loading density or theoretical maximum density of an explosive ($P_{det,max}$) with the general formula $C_aH_bN_cO_d$ can be given as follows [93]:

$$
\begin{aligned}
P_{det,max} = {} & 221.5 - 20.44a - 2.254b + 17.22c + 16.14d \\
& - 79.07C_{SSP} - 66.34n_N,
\end{aligned} \tag{4.10}
$$

where $P_{det,max,SSP}$ is equal to 1.0 for explosives which contain N=N−, −ONO$_2$, NH$_4^+$, or −N$_3$ in the molecular structure; n_N is equal to $0.5n_{NO_2} + 1.5$ where n_{NO_2} is the number of nitro groups attached to carbon in nitrocompounds in which $a = l$. For example, applying equation (4.10) for cyclotol-50/50, with the composition 50/50 RDX/TNT ($C_{2.22}H_{2.45}N_{2.01}O_{2.67}$), results in $P_{det,max}$ as

$$
\begin{aligned}
P_{det,max} = {} & 221.5 - 20.44 \times 2.22 - 2.254 \times 2.45 + 17.22 \times 2.01 + 16.14 \times 2.67 \\
& - 79.07 \times 0 - 66.34 \times 0 \\
= {} & 248.3\ \text{kbar.}
\end{aligned}
$$

The measured value of the detonation pressure for Cyclotol-50/50 at a maximum loading density of 1.63 g/cm^3 is 231 kbar [39].

4.4 Prediction of the detonation pressure of nonideal aluminized explosives

It was shown that the calculated C–J detonation parameters of nonideal explosives obtained from using existing thermodynamic computer codes are significantly different from experimental results [39]. It can be assumed that nonequilibrium effects in the reaction zones may contribute to this deviation. The measured detonation pressures may be higher than equilibrium calculations if the measurement is taken behind the von-Neumann spike but in front of the C–J point [39]. For aluminized explosives, the mean size of aluminum particles used is around 101 μm. However, aluminum powder needs to be excited for several μs before it participates in the chemical reaction, but the reaction time of the reaction zones is about 10^{-1} μs [94]. Therefore, it is difficult for aluminum powder to participate in the chemical reaction at the reaction zones. The detonation pressure can be predicted through use of a computer code and EOS by means of the C–J thermodynamic detonation theory, which assumes that the thermodynamic equilibrium is reached instantaneously. During the expansion of the gaseous detonation products, the combustion of aluminum particles in explosives is assumed to occur behind the reaction front [94]. Thus, aluminum particles in this case do not participate in the reaction zone and instead act as inert ingredients [12, 39].

Zhang and Chang [94] have suggested that the best agreement with the experimental data for aluminized explosives was obtained by adjusting parameter κ in the BKW-EOS, i. e. equation (1.13). They argued that the value of κ depends on the fraction of solid products in the C–J reaction, and it should be adjusted when they increase. The adjusted values of κ for RDX-type and TNT-type explosives in the BKW-EOS are 9.2725 and 10.4017, respectively. Using this modification, they have calculated detonation pressures and velocities of aluminized explosives within about 9 % and 7 % of the experimentally obtained values respectively.

Besides complex thermochemical computer codes, there are several empirical methods which can be used for calculating the detonation pressure of aluminized explosives, and which are discussed in the following sections.

4.4.1 Using the elemental composition for predicting the detonation pressure of explosives

For aluminized explosives with the general formula $C_a H_b N_c O_d Al_e$, it was shown that the elemental composition and loading density are sufficient for predicting the detonation pressure according to [30]:

$$P_{det} = -35.53a + 41.42b - 14.77c + 44.00d - 21.32e + 43.95\rho_0^2. \qquad (4.11)$$

This correlation is limited to explosives containing aluminum, and it cannot be used for pure and composite $C_a H_b N_c O_d$ explosives. For example, the value for the detonation pressure of Alex 20 with composition 44/32/20/4 RDX/TNT/Al/Wax ($C_{1.783}H_{2.469}N_{1.613}O_{2.039}Al_{0.7335}$) at a loading density of 1.801 g/cm^3 is calculated as follows:

$$P_{det} = -35.53 \times 1.783 + 41.42 \times 2.469 - 14.77 \times 1.613 + 44.00 \times 2.039$$
$$- 21.32 \times 0.7335 + 43.95(1.801)^2$$
$$= 231.8 \text{ kbar.}$$

The measured detonation pressure for Alex 20 is 230 kbar [12].

4.4.2 Detonation pressure of $C_aH_bN_cO_dF_eCl_f$ and aluminized explosives as a function of the loading density, element composition, and the condensed phase heat of formation of pure or composite explosives

For aluminized explosives, it is possible to improve the decomposition paths given in equation (1.11) to calculate the detonation pressure of aluminized explosives with the general formula $C_aH_bN_cO_dF_eCl_fAl_g$ as follows [28]:

$$C_aH_bN_cO_dF_eCl_fAl_g \longrightarrow e\,HF + f\,HCl(g) + \tfrac{c}{2}N_2(g)$$

$$\begin{cases}
(d-0.15g)CO(g) + (a-d+0.15g)C(s) + (\tfrac{b-e-f}{2})H_2(g) \\
+ 0.05g\,Al_2O_3(s) + 0.9g\,Al(s) \\
\text{with } d \leq a & \text{(a)} \\[6pt]

a\,CO(g) + (d-a-0.25g)H_2O + (\tfrac{b-e-f}{2} - d + a + 0.25g)H_2(g) \\
+ 0.125g\,Al_2O_3(s) + 0.75g\,Al(s) \\
\text{with } d > a,\ \tfrac{b}{2} > d-a,\ d-a \geq 0.25g & \text{(b)} \\[6pt]

(d-0.375g)CO(g) + (\tfrac{b-e-f}{2})H_2(g) + (a-d+0.375d)C(s) \\
+ 0.125g\,Al_2O_3(s) + 0.75g\,Al(s) \\
\text{with } d > a,\ \tfrac{b}{2} > d-a,\ d-a < 0.25g & \text{(c)} \\[6pt]

(\tfrac{b-e-f}{2} - 0.25g)H_2O(g) + 0.25g\,H_2(g) \\
+ (2a-d+\tfrac{b-e-f}{2})CO(g) + (d-a-\tfrac{b-e-f}{2})CO_2(g) \\
+ 0.125g\,Al_2O_3(s) + 0.75g\,Al(s) \\
\text{with } d \geq a + \tfrac{b-e-f}{2},\ d \leq 2a + \tfrac{b-e-f}{2},\ \tfrac{b-e-f}{2} \geq 0.25g & \text{(d)} \\[6pt]

(\tfrac{b-e-f}{2})H_2(g) + (2a-d+b-e-f)CO(g) \\
+ (d-a-\tfrac{b-e-f}{2} - 0.1875)CO_2(g) \\
+ 0.125g\,Al_2O_3(s) + 0.5g\,Al(s) \\
\text{with } d \geq a + \tfrac{b-e-f}{2},\ d \leq 2a + \tfrac{b-e-f}{2},\ \tfrac{b-e-f}{2} < 0.25g & \text{(e)} \\[6pt]

(\tfrac{b-e-f}{2})H_2O(g) + a\,CO_2(g) + (\tfrac{2d-b+e+f}{4} - a - 0.375)O_2(g) \\
+ 0.25g\,Al_2O_3(s) + 0.5g\,Al(s) \\
\text{with } d \geq 2a + \tfrac{b-e-f}{2},\ \tfrac{2d-b+e+f}{4} - a \geq 0.375g & \text{(f)}
\end{cases}$$

(4.12)

It was shown that the following equation can be used to predict the detonation pressure of ideal and non-ideal aluminized explosives, based on above decomposition paths as [28]:

$$P_{det} = 252.8(n'_{gas})(\bar{M}_{wgas}Q_{det}[H_2O(g)])^{0.5}\rho_0^2 - 14.84. \tag{4.13}$$

Equation (4.13) is an improved correlation of equation (4.7), and can be applied for not only ideal $C_aH_bN_cO_dF_eCl_f$ explosives, but also for nonideal aluminized explosives. For example, as is shown in the Appendix, RDX/Al (90/10) with the empirical formula $C_{1.215}H_{2.43}N_{2.43}O_{2.43}Al_{0.371}$ has $\Delta_f H^\theta(c) = 24.89$ kJ/mol. Therefore, the values of n'_{gas}, $\bar{M}_{wgas}$, and $Q_{det}[H_2O(g)]$ are 0.0365 mol/g, 24.06 g/mol, and 4.97 kJ/g respectively. The use of these values in equation (4.13) gives the detonation pressure of RDX/Al (90/10) at a loading density of 1.68 g/cm³ as

$$P_{det} = 252.8(0.0365)(24.06 \times 4.97)^{0.5}(1.68)^2 - 14.84$$
$$= 270 \text{ kbar,}$$

which can be compared with the measured value of RDX/Al (90/10) which is 246 kbar [39].

4.4.3 Using molecular structure for predicting the detonation pressure of ideal and aluminized explosives

It was found that the detonation pressure of ideal $C_aH_bN_cO_d$ explosives and aluminized explosives can be predicted using [33]

$$P_{det} = -23.35 + 105.9\rho_0^2 - 12.39a - 1.83b + 6.50c + 5.40d$$
$$- 24.71n_{NR_1R_2} - 63.08n'_{Al}, \tag{4.14}$$

where $n_{NR_1R_2}$ is the number of $-NH_2$, NH_4^+, and five membered rings with three (or four) nitrogens in any explosive, as well as five (or six) membered rings in nitramine cages; n'_{Al} is a function of the number of moles of Al which can be determined according to the following conditions:
(i) $n'_{Al} = 1.5n_{Al}$ for $d \le a$;
(ii) $n'_{Al} = 1.4n_{Al}$ for $d > a + \frac{b}{2}$ and $d \le 2a + \frac{b}{2}$;
(iii) $n'_{Al} = 1.25n_{Al}$ for $d > a$ and $\frac{b}{2} \ge d - a$;
(iv) $n'_{Al} = n_{Al}$ for $d > 2a + \frac{b}{2}$.

If the mass ratio of aluminum to explosive $\ge \frac{2}{3}$, or $(\frac{m_{Al}}{m_{CHNO}}) \ge \frac{2}{3}$, n_{Al} in the above conditions should be multiplied by 0.6.

To use equation (4.11) for aluminized explosives, 100 g of explosives with the general formula $C_aH_bN_cO_dAl_e$ were used for the calculation of the detonation pressure, and the number of moles of extra aluminum in explosives, e. g. RDX/Al (50/50) has the empirical formula $C_{0.675}H_{1.35}N_{1.35}O_{1.35}Al_{1.853}$. However, for a 100 g mixture of RDX and Al, the formula should be changed to $C_{1.35}H_{2.70}N_{2.70}O_{2.70}Al_{3.706}$, i. e.

$$C: \quad 0.675 \times \frac{100}{50} = 1.35;$$

$$H: \quad 1.35 \times \frac{100}{50} = 2.70;$$

$$N: \quad 1.35 \times \frac{100}{50} = 2.70;$$

$$O: \quad 1.35 \times \frac{100}{50} = 2.70;$$

$$Al: \quad 1.853 \times \frac{100}{50} = 3.706.$$

Since RDX/Al (50/50) follows condition (iii) as well as $(\frac{m_{Al}}{m_{CHNO}}) \geq \frac{2}{3}$, its detonation pressure at loading density 1.89 g/cm^3 is calculated as follows:

$$
\begin{aligned}
P_{det} = \ & -23.35 + 105.9(1.89)^2 - 12.39 \times 1.35 \\
& - 1.83 \times 2.70 + 6.50 \times 2.70 + 5.40 \times 2.70 \\
& - 24.71 \times 0 - 63.08 \times 3.706 \times 1.25 \times 0.6 \\
= \ & 190 \, \text{kbar}.
\end{aligned}
$$

The predicted value is the same as the measured value of 190 kbar [39].

Deviations in the values calculated using the new method from the experimental values are large for two cases:

(i) if $d > 3(a + b)$, e. g. TNM;

(ii) if the mass percent of nonenergetic additives is large.

4.4.4 Maximum attainable detonation pressure of $C_aH_bN_cO_dF_e$ explosives and aluminized explosives

For $C_aH_bN_cO_dF_e$ explosives and aluminized explosives, it was shown that the following equation can be used to predict the maximum attainable detonation pressure ($P_{det,max}$) of these explosives [34]:

$$
\begin{aligned}
P_{det,max} = \ & 216 - 13.9a - 3.30b + 18.1c + 5.88d \\
& + 101P'_{in} - 68.0P'_{de},
\end{aligned}
\tag{4.15}
$$

where P'_{in} and P'_{de} are two correcting functions, which are specified as follows.

4.4.4.1 Pure $C_aH_bN_cO_dF_e$ explosives

(i) P'_{in}: The value of P'_{in} equals 1.1 for CNO explosives without azido groups.
(ii) P'_{de}: Two different categories of compounds may need this correcting function:

(a) if $a = 1$, $P'_{de} = 1.1n_{NO_2}$ where n_{NO_2} is the number of nitro groups,

(b) for the presence of $-N=N-$ in non-aromatic (or between two aromatic rings) compounds, ether, oxamide, and non-aromatic fluorinated nitramines, $P'_{de} = 0.7$.

The two correcting functions P'_{in} and P'_{de} may exist simultaneously, e. g. the values of P'_{in} and P'_{de} in TNM are 1.1 and 4.4, respectively.

4.4.4.2 Mixture of $C_aH_bN_cO_dF_e$ and aluminized explosives

All calculations are based on a 100 g mixture of different components.

(i) $C_aH_bN_cO_dF_e$ explosives:

(a) For a mixture of TNT and nitramine, the values of P'_{in} and P'_{de} are the mass fractions of nitramine and TNT, respectively;

(b) the value of P'_{in} is 0.7 for plastic bonded explosives;

(c) for a liquid mixture of explosives in which one explosive component contains one carbon atom such as NM, the value of P'_{de} is equal to 1.2.

(ii) Aluminized explosives: If the mass percent of aluminum is greater than or equal to 30, the value of P'_{de} is equal to 0.5.

For example, for a 74.766/18.691/4.672/1.869 TNT/Al/Wax/Graphite mixture with general formula $C_{2.79}H_{2.31}N_{0.99}O_{1.98}Al_{0.69}$, the value of $P_{det,max}$ is given as follows:

$$P_{det,max} = 216 - 13.9 \times 2.79 - 3.30 \times 2.31 + 18.1 \times 0.99 + 5.88 \times 1.98$$
$$+ 101 \times 0 - 68.0 \times 0$$
$$= 199 \text{ kbar}.$$

The measured detonation pressure of this composite explosive at a loading density of 1.68 g/cm^3 is 175 kbar [12].

4.5 Application of laser techniques for assessment of the detonation performance

Since new energetic compounds are synthesized in milligram quantities initially, their actual scale-up to test the detonation performance requires significant quantity of energetic materials. Many interesting candidate energetic materials are introduced each year. Meanwhile, there are limited resources for their scale-up and detonation testing. Thus, it is essential to have methods for down-selecting the most promising materials.

It was indicated in recent years that it is possible to use laser techniques for assessment of the detonation velocities and detonation pressures of energetic materials prior to their scale-up and conventional detonation testing [35, 95–99]. Energetic materials do not detonate because of excitation with the nanosecond-pulsed laser. Thus, the results presume that the material is detonable and that the reactions in the plasma are comparable to those behind the detonation front. Confirmation of the results of laser techniques in detonation testing is essential but laser techniques provide valuable information about the material that is not otherwise obtainable when only milligram quantities are available.

If a nanosecond-pulsed laser with sufficient energy is focused onto a sample surface, the material is ablated and a laser-induced plasma is formed where it atomizes, ionizes, and excites the ablated material. Properties of plasma depend on the material and the laser parameters including pulse width, energy and wavelength. After attaining local thermodynamic equilibrium, typical temperatures are in the range from around 5 000–20 000 K [95].

4.5.1 The laser-induced air shock from energetic materials (LASEM) method

A laboratory-scale method has been developed recently to measure the rapid energy release from milligram quantities of energetic material based on the high-temperature chemistry, which is induced by a focused nanosecond laser pulse. A high-speed camera is used to record the expansion of the shock wave into the air above the sample surface, which is due to ensue increment of exothermic chemical reactions result in velocity of the laser-induced shock wave. There is a strong linear correlation between the laser-induced shock velocity and the reported detonation velocities from the large-scale detonation testing. The **laser-induced air shock from energetic materials (LASEM)** method can be introduced as a means of estimating the detonation performance of novel energetic materials prior to their scale-up and full detonation testing [95–97]. LASEM method can bridge the knowledge gap between conventional analytical techniques available to characterize milligram quantities of material, such as the bomb calorimetry and **differential scanning calorimetry (DSC)** , and the full-scale detonation testing [100, 101]. It is based on the high-temperature chemical reactions of a material in a laser-induced plasma. The released energy affects the resulting laser-induced shock wave that expands into the air above the sample. Thus, it can be expected that the high-temperature chemical reactions of energetic materials in the laser-induced plasma liberate enough energy to influence the expansion of the laser-induced shock wave produced by the plasma formation. The release of additional energy from the reactions in the plasma slows the deceleration of the shock wave as it expands into the air. Exothermic reactions occurring in the laser-induced plasma can rise the plasma temperature [101, 102] and, in underwater applications, increase the bubble energy [103].

There are strong correlations between the CN/C_2 emission ratios from the laser-induced plasmas of a variety of novel energetic materials with their predicted detonation velocity and pressure [104]. Plasma thrusters or laser chemical reaction engines may be used as the laser ablation of energetic polymers for propulsion applications [105]. It was found that the exothermic reactions of laser ablated energetic materials increased the velocities of laser-induced shock waves as they expanded into the air above the residue sample surface [100].

The laser ablation of energetic materials can provide large shock velocities (>750 m/s) that may produce significantly larger heat-affected zones in the surrounding atmosphere. Laser-induced deflagration reactions can occur on the millisecond timescale as material was ejected from the sample surface into the heat-affected zone following the passage of the shock wave. They can characterize the slow energy release from energetic materials during their combustion with air [106]. Figure 4.3 shows the LASEM experimental setup, consists of a 1064-nm, 900-mJ, 6-ns pulsed Nd:YAG laser focused just below the residue sample surface to minimize entrainment of air into the laser-induced plasma [95–97].

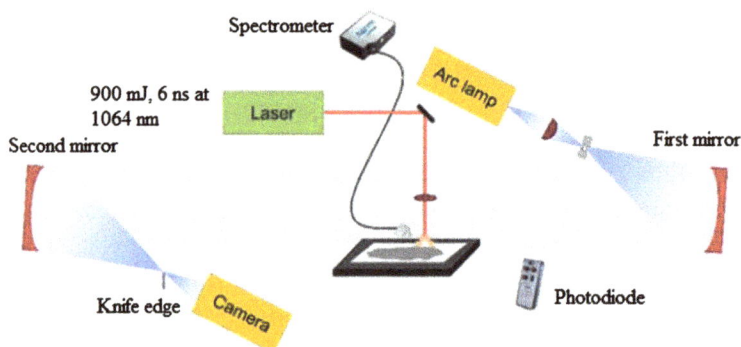

Figure 4.3: Experimental layout for direct laser excitation of energetic materials [95–97].

4.5.2 Application of LASEM for composite energetic materials with metal additives

The method of LASEM has been used to estimate detonation velocities from RDX+Al and TNT+Al explosive formulations compared with the measured detonation velocities reported in the literature [107]. The measured laser-induced shock velocities for the metal-containing formulations can be assessed by the calibration curve relating the laser-induced shock velocities to the measured detonation velocities of organic military explosives. Figure 4.4 shows the correlation between laser-induced shock velocities and the estimated detonation velocities for conventional military explosives with metal additives [95]. As seen, there is a linear relationship between the estimated detonation velocities for the formulations agreed with the average detonation velocities from the large-scale detonation testing. For formulation TNT/Al (80/20), containing 20 % Al (by weight) to TNT, the estimated detonation velocity is 6.88 ± 0.28 km/s, which is lower than pure TNT (7.14 ± 0.30 km/s) and close to the corrected value of the average detonation velocity at the theoretical maximum density, i. e. 6.82 ± 0.16 km/s.

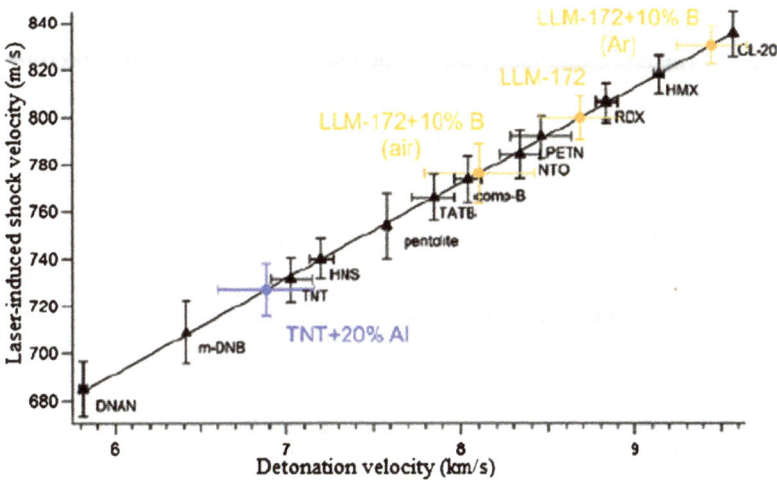

Figure 4.4: Laser-induced shock velocities and the estimated detonation velocities for some metalized explosives [95].

Application of boron (B) with the highest volumetric heat of oxidation (140 GJ m^{-3}) may be suitable for a hydrogen-less explosive because its use in composite energetic materials has been hindered by the formation of HBO_2. The product HBO_2 is an energetically less favorable one that prevents complete oxidation to B_2O_3. Figure 4.4 shows the addition of 10 % B to the hydrogen-less explosive 3,4-bis(4-nitro-1,2,5-oxadiazol-3-yl)-1,2,5-oxadiazole (LLM-172) can decrease the estimated detonation velocity compared to pure LLM-172. Meanwhile, the laser-induced plasma with Ar rather than air shows a substantial increase in the expected detonation performance for the formulation. In Figure 4.4, the y-error bars are 95 % confidence intervals (C. I.) for the laser-induced shock velocities, while the x-error bars for the conventional energetic materials (black) are 95 % C. I. for the reported detonation velocities. For the composites with metal additives (blue, orange), x-error bars are 95 % C. I. for the estimated detonation velocities from LASEM.

4.5.3 Laser Induced Breakdown Spectroscopy (LIBS)

Laser Induced Breakdown Spectroscopy (LIBS) is introduced as a successful analytical method for elemental analysis [108, 109]. It has been developed to investigate and analyze the energetic materials [110, 111]. It uses the light emitted from laser-generated microplasma for determination of the composition of the sample based on elemental and molecular emission intensities. It has superior advantages including non-destructive, simultaneous multi-elemental analysis, fast detection, no initial preparation, online and in situ detection [112, 113]. For the study of spectroscopy signatures of molecular compounds, the selected diatomic molecular emissions have also been applied [114]. The molecular emission can be related to its molecular structure, which can yield a deeper understanding of the principal formation routes of molecular species existing in laser-induced plasmas [115]. Molecular components may be formed from the ablated material. Excited atoms can be recombined with ions or with air components in the laser-induced plasma. For aluminized energetic compounds, it was found that two molecular species, AlO and CN, are observed in a like manner [35, 113, 116]. The statistics of full spectra of organic energetic materials can be used for their identification specifically because spectral analysis aims either at extracting qualitative information for identification/classification purposes or at calculating concentrations [117]. The LIBS method may utilize linear or parametric correlation, **principle component analysis (PCA)**, and **partial least squares discriminant analysis (PLS-DA)** to identify organic compounds [118]. PCA can discriminate between two classes of spectra, and also detect outliers. It can visualize similarities between spectra through graphically representing them in a new space with much fewer dimensions. This situation may retain most of the information contained in the data set [119, 120].

For quantitative analysis, the calibration curve method is one of the most popular methods among several different ways to determine the concentration. A curve is drawn in this method between the integrated intensity of the atomic emission of the element and the corresponding concentration of that element in the standard samples. This calibration curve can be used to determine the concentration of the element in the unknown sample by measuring the intensity of the corresponding atomic emission in the LIBS spectra of the sample. This method is most practical because the standard or certified reference material (CRM) is easily available to extract quantitative information about the constituents of the sample [121–123]. Figure 4.5 shows the schematic diagram of the LIBS set-up, which has been used to determine the detonation performance of RDX-based aluminized explosives, aluminized TNT explosives and the aluminized plastic-bonded explosive (PBX).

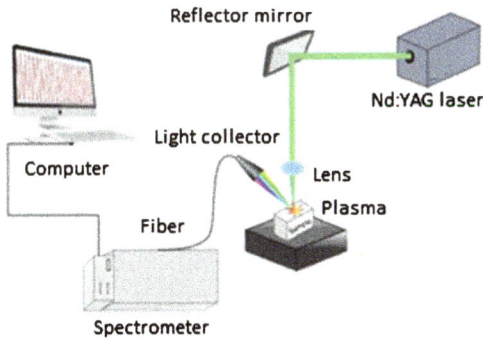

Figure 4.5: Schematic diagram of the experimental LIBS set-up.

4.5.3.1 Detonation performance of RDX-based aluminized explosives

Five different compositions of RDX-based aluminized explosives including RDX/Al (90/10), RDX/Al (80/20), RDX/Al (70/30), RDX/Al (60/40) and RDX/Al (50/50) as standard samples were used to study their detonation performance by the LIBS method [35]. The LIBS measurements were done in two different air and argon atmospheres.

As indicated, the detonation velocity and pressure of non-ideal aluminized explosives depend on different parameters such as elemental composition, heat of formation and the initial density. For five compositions of RDX-based aluminized explosives, the measured or reliable calculated detonation parameters including velocity [124], pressure [124], temperature [125] and heat [125] and the Al/O composition ratio are given in Table 4.1.

Table 4.1: The measured or reliable calculated detonation parameters and the Al/O composition ratio for five samples of the RDX/Al.

Sample	Detonation velocity [124] (km/s)	Detonation pressure [124] (kbar)	Heat of detonation [125] (kJ/kg)	Temperature of detonation [125] (K)	Al/O ratio
RDX/Al (90/10)	8.02	245	6069	3600	0.15
RDX/Al (80/20)	7.77	230	7904	4972	0.33
RDX/Al (70/30)	7.51	217	9611	5447	0.59
RDX/Al (60/40)	7.23	203	8204	4017	0.92
RDX/Al (50/50)	6.95	189	6673	2316	1.37

Equation (4.16) shows a good linear relationship ($r^2 > 0.95$), where r^2 is the correlation of determination, between the Al/O intensity ratio and the Al/O composition ratio in five samples of the RDX/Al given in Table 4.1 under Ar atmosphere. It was indicated that the relative intensities of the Al/O from LIBS measurements can be linearly correlated with the detonation velocity and pressure of RDX/Al standard samples. Equations (4.17) and (4.18) show a good linear correlation between the Al/O intensity ratio and the detonation velocity/pressure in the RDX/Al standard samples ($r^2 > 0.99$) [35]. Thus, these equations can be used to determine the detonation velocity and pressure of the unknown samples containing RDX/Al [35].

$$\left(\frac{n_{Al}}{n_O}\right)_{LIBS} = 36.17 \times \left(\frac{n_{Al}}{n_O}\right)_{theory} + 4.769, \tag{4.16}$$

$$D_{det} = -0.02485 \times \left(\frac{n_{Al}}{n_O}\right)_{LIBS} + 8.205, \tag{4.17}$$

$$P_{det} = -1.291 \times \left(\frac{n_{Al}}{n_O}\right)_{LIBS} + 253.5, \tag{4.18}$$

where $(\frac{n_{Al}}{n_O})_{LIBS}$ and $(\frac{n_{Al}}{n_O})_{theory}$ are the Al/O intensity ratios, which were determined and calculated by the LIBS method and chemical composition, respectively. For deriving these equations, the intensity ratio of two atomic lines (Al/O) rather than the Al line intensity has been used in order to reduce the standard deviation and the influence of experimental parameters like laser power, sample-to-lens distance, and the nature of the matrix elements of different samples. It should be mentioned that equations (4.16) to (4.18) were obtained from the calibration curves, which are only valid for RDX/Al mixtures in the same %Al range (10–50 %). Since the compositions of RDX/Al were prepared from the military grade RDX (<800 μm particle diameter) and aluminum powder (typically ~50 μm average diameter), it can be expected that different calibration curves may be obtained using different particle sizes and adding binders, e. g. different plasma chemistries.

Figure 4.6 shows graphs of the ratio of intensities of molecular bands (AlO) to sum of intensities of atomic lines (Al+O) in argon atmosphere as well as the measured and calculated values of heat and temperature of explosion [125] versus the aluminum content of RDX/Al standard samples. As indicated in Figure 4.6, there is a steep decrease in the ratio of AlO/(Al+O) by increasing aluminum content up to 30 % and then fixed. Both the heat and temperature of explosion are increased with the aluminum content up to 30 % and then strongly decreased by increment of the aluminum [35].

Thus, increasing aluminum composition percentage up to 30 % can increase the oxidation of Al to form Al_2O_3, which rise temperature and heat of explosion. Moreover, the formation of Al+O is preferred instead of the formation of molecular bond AlO within the plasma. It can be concluded that the released energy in the laser-induced plasma of the RDX/Al standard samples is related to the heat of explosion and the aluminum content.

4.5.3.2 Detonation performance of TNT-based aluminized explosives

For four different compositions of aluminized TNT, the measured density, the detonation parameters [126] and the Al/O composition ratio are given in Table 4.2. The test sample charge size was 22 mm Ø × 90 mm [126].

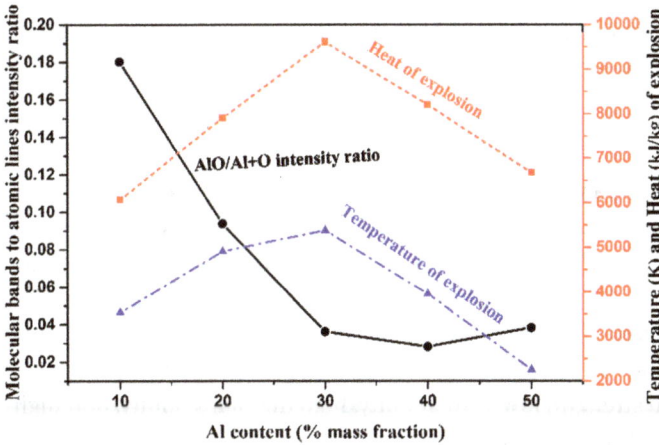

Figure 4.6: The intensity ratio AlO/Al+O (circle) as well as experimental values of temperature (triangle) and heat (square) of explosion versus percent of mass fraction aluminum in the RDX/Al standard samples [35].

Table 4.2: Composition, explosive properties and Al/O composition ratio for four TNT/Al standard samples.

Composition	Experimental density (g cm^{-3})	Detonation velocity (km/s)	Detonation pressure (kbar)	Al/O ratio
TNT/Al (95/5)	1.63	6.829	190	0.07
TNT/Al (90/10)	1.66	6.659	184	0.15
TNT/Al (85/15)	1.68	6.528	179	0.25
TNT/Al (80/20)	1.72	6.325	172	0.35

It was found that similar linear relationships, as those given for RDX/Al compositions, i. e. equations (4.19) to (4.21), were obtained for TNT/Al compositions as follows [99]:

$$\left(\frac{n_{Al}}{n_O}\right)_{LIBS} = 134.3 \times \left(\frac{n_{Al}}{n_O}\right)_{theory} + 16.75, \tag{4.19}$$

$$D_{det} = -0.01344 \times \left(\frac{n_{Al}}{n_O}\right)_{LIBS} + 7.185, \tag{4.20}$$

$$P_{det} = -0.481 \times \left(\frac{n_{Al}}{n_O}\right)_{LIBS} + 202.7, \tag{4.21}$$

equation (4.19) shows a good linear relationship ($r^2 > 0.99$) between the Al/O intensity ratio and the Al/O composition ratio in the TNT/Al standard samples under an Ar atmosphere. Equations (4.20) and (4.21) indicate good linear correlations between the

Al/O intensity ratio and detonation velocity/pressure in the TNT/Al standard samples ($r^2 > 0.99$) [99]. Thus, they can determine the detonation velocity and pressure of an unknown TNT/Al sample.

4.5.3.3 Detonation performance of PBX

Since TNT-based explosives have poor mechanical properties, defects due to shrinkage, low thermal stability, and high sensitivity to different types of stimuli, these drawbacks can be improved, to a large extent, in plastic-bonded castable compositions. The aluminized **plastic-bonded explosive (PBX)** includes three major components, such as polymeric binder, metal fuel, and nitramine explosive. The PBX system is characterized by high mechanical strength, low vulnerability, high thermal stability, and high-energy output. The **hydroxy-terminated polybutadiene (HTPB)** is usually used in PBXs because it has high-temperature resistance with the temperature range from −54 to +71 °C and low glass transition temperature (−79 °C). For manufacturing PBXs, the HTPB is used with the other components such as bonding agent, coupling agent, plasticizer, processing aid, and curing agent, which make the mixture blending with good mechanical and physical properties.

The detonation velocity and pressure of aluminized PBXs are directly related to the elemental composition, the heat of formation and the initial density [28]. The composition, the measured density, the **theoretical maximum density (TMD)** , and the detonation velocity and pressure [127] for five standard samples of RDX/Al/HTPB are shown in Table 4.3.

Table 4.3: Composition and explosive properties for five standard samples of RDX/Al/HTPB [127].

RDX/Al/HTPB	TMD (g/cc)	Experimental density (g/cc)	Detonation velocity (km/s)	Detonation pressure (kbar)
80/5/15	1.609	1.594	7.530	229
75/10/15	1.630	1.610	7.500	223
70/15/15	1.670	1.630	7.580	224
65/20/15	1.680	1.646	7.257	220
60/25/15	1.709	1.680	7.108	218

In contrast to binary compositions RDX/Al and TNT/Al, there are no linear correlations as those given by equations (4.19) to (4.21) for PBX samples. Thus, the other LIBS data were investigated to find suitable correlations. Figure 4.7 shows the intensity ratio of CN (388.29 nm)/C (247.8 nm) in an argon atmosphere as well as the experimental values of velocity and pressure versus the aluminum content of RDX/Al/HTPB, which are given in Table 4.3 [98]. As shown in Figure 4.7, the detonation velocity and pressure de-

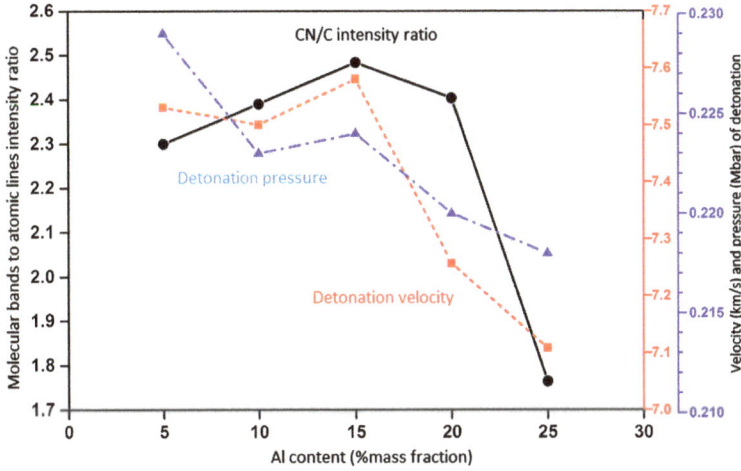

Figure 4.7: Molecular bands to atomic lines intensity ratio (circle) as well as experimental values of detonation velocity (square) and pressure (triangle) [127] versus aluminum content of five samples of RDX/Al/HTPB in argon atmosphere [98].

crease gradually with increasing the aluminum content up to 10 percent (unlike CN/C intensity ratio), followed by an increase up to 15 percent (like CN/C intensity ratio). For a higher Al content, the detonation parameters and the CN/C intensity ratio decrease steadily.

The intensity variation of the CN emission is a function of the nitrogen contents of RDX of the studied samples. Thus, the CN emission intensity increases and decreases as the mole fraction of both carbon and nitrogen increase and decrease, respectively [128]. The results suggest that the intensity ratio of CN/C is related to the detonation performance and the aluminum content.

4.6 Calculating the detonation pressure of ideal and non-ideal explosives containing Al and AN

It was recently indicated that it is possible to use reaction paths of equation (1.23) for assessment of the detonation pressure of non-ideal explosives containing Al and AN as follows [36]:

$$P_{det} = 244.36(n'_{gas})(\bar{M}_{wgas}Q_{det}[H_2O(g)])^{0.5}\rho_0^2 - 8.74. \tag{4.22}$$

For example, as was indicated in Table 3.3, the values of n'_{gas}, $\bar{M}_{wgas}$ and $Q_{det}[H_2O(g)]$ for LX-17 are 0.0444 mol/g, 22.12 g/mol and 1.95 kJ/g, respectively. For

$\rho_0 = 1.91\,\text{g/cm}^3$, the value of the calculated detonation pressure is 251 kbar where the reported detonation pressure for LX-17 at this ρ_0 is 260 kbar [129].

Summary

In contrast to available methods for prediction of the detonation velocity, there are fewer predictive methods for the detonation pressure. Two reasons may contribute to this situation:

(a) there is only small amount of experimental data for detonation pressures,
(b) there is a large uncertainty in the reported detonation pressures because of the existence of nonequilibrium effects in reaction zones.

The best available predictive methods were reviewed in this section for both ideal and nonideal aluminized explosives. The methods which have been described can be applied for the design of new pure and composite explosives without requiring the use of complex thermochemical computer codes.

Two techniques of LASEM and LIBS have also been introduced as two new methods for assessment of the detonation performance of non-ideal aluminized explosives.

4.7 Questions and problems

i The necessary information for some problems are given in the Appendix.

(1) (a) Calculate n'_{gas}, $\bar{M}_{\text{wgas}}$, and $Q_{\text{det}}[H_2O(g)]$ for RDX/TFNA (65/35) using equation (4.7).
     (b) If the measured detonation pressure of RDX/TFNA (65/35) at loading density $1.754\,\text{g/cm}^3$ is 324 kbar, calculate its detonation pressure and percent deviation by equation (4.7).
(2) If the gas phase heat of formation of Cyclotol-78/22 is 12.88 kJ/mol, use equation (4.8) to calculate its detonation pressure at loading density $1.76\,\text{g/cm}^3$.
(3) Use equation (4.9) to calculate the detonation pressure of BTF at loading density $1.76\,\text{g/cm}^3$.
(4) Use equation (4.10) to calculate the maximum attainable detonation pressure of tetranitromethane (TNM).
(5) Using equation (4.11), calculate the detonation pressure of TNT/Al (78.3/21.7) at loading density $1.80\,\text{g/cm}^3$.
(6) Use equation (4.13) to calculate the detonation pressure of RDX/Al (50/50) at loading density $1.89\,\text{g/cm}^3$.

(7) Calculate the maximum attainable detonation pressure of the following molecular structure using equation (4.14).

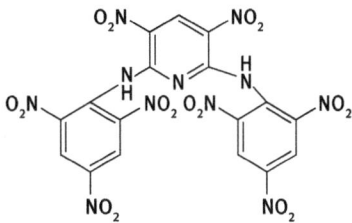

(8) Use equation (4.22) to calculate the detonation pressure of RDX/Al (60/40) at loading density 1.84 g/cm^3 and percent of its deviation with the reported value, i. e. 211 kbar [42].

For answers and solutions, please see p. 120.

5 Gurney energy

High explosives are an important part of many types of ammunition because the damage inflicted on a given target is the result of blast and fragmentation from the detonation of the high explosive. There are three categories for fragmentation warheads including natural fragmentation, controlled fragmentation and preformed fragments [130]. Natural fragmentation warheads can give fragments exhibiting a wide range of size and shape of poor lethality. This situation may be due to the fact that nonoptimized irregular-shaped fragments tend to experience high aerodynamic drag forces when flying through the air, resulting in considerable velocity loss [131]. For controlled fragmentation and preformed fragments, introducing stress raisers such as grooves on the inside surface of the casing, or by embedding fragments of different geometries between the explosive charge and the metal casing can produce fragments of particular shape, mass and number.

Gurney [132] developed a simple model for the estimation of the velocity of the surrounding layer of metal (or other material) when an explosive detonates. The **specific energy** or **Gurney energy** (E_G) is more useful than detonation properties for the ballistic characterization of an explosive, because it can permit calculation of the velocities and impulses imparted to metal fragments driven by detonating high explosives [82]. Thus, a given explosive liberates a fixed amount of E_G on detonation, which is converted to kinetic energy and transferred to the moving metal fragments and gaseous products. The value of E_G determines the amount of mechanical work that can be produced by the explosive and which is necessary for acceleration of the surrounding metal.

The internal energy during expansion of the detonation products from the C–J point can be related to any measure of the energy output from a detonating explosive. The process of metal acceleration has a limited duration. The actual amount of energy generated by the explosive which used to accelerate metal fragments is less than the energy of the explosive as determined by detonation calorimetry, because the internal energy remaining in the gaseous products on further expansion does not contribute to the metal acceleration. **Gurney's model** can determine only the final metal velocity and gives no information about the acceleration process. Acceleration of the tube wall stops soon after the wall is strained to the point of rupture, because the gases leak out between fragments in divergent geometries. Thus, the value of E_G is only a fraction of the stored chemical energy in the initial explosive mixture, because the detonation products are still hot in expansion to ambient pressure. The measured E_G is significantly less than the chemical energy of the unreacted explosive [82]. The value of E_G is a function of the chemical energy and the density of an explosive [82]. Gurney's model underestimates metal velocities when $\frac{m}{c} < \frac{1}{3}$, where m and c are the masses per unit length of the metal and the explosive respectively [82]. In such situations the efficiency of the explosive is increased and shock processes associated with the deto-

https://doi.org/10.1515/9783110677652-005

nation front dominate the acceleration behavior. Thus, a gas dynamic model or wave propagation computations should be used at low values of $\frac{m}{c}$.

5.1 Gurney energy and Gurney velocity

The Gurney model can be used to determine the amount of chemical energy which is transformed into kinetic energy for the expanding product mixture and moving metal fragments by calculating the difference between the internal energy of the **isentropically expanded products** (U_s) and that of the **unreacted explosive** (U_0). This energy difference is assumed to be a measure of the energy which is available for accelerating the metal, and ignores the wave dynamics in the flow of the product gases. For Gurney's model all fragments are assumed to be releases at the same initial velocity, where the velocity of the gaseous explosion products increases from zero at the center of mass of the explosive out to a maximum, which is also the velocity of the casing fragments at the moment of break-up. The **Gurney velocity** or **Gurney constant** ($\sqrt{2E_G}$) is related to the value of E_G, which provides a more relevant absolute indicator of the ability of an explosive to accelerate metal under a wide variety of loading conditions and geometries. Thus, the Gurney model permits a quantitative estimation of the velocity or impulse imparted to metal by detonating explosives, rather than just simply a rank-ordering of the explosives. For this model, the velocity profile of the product gases is also linear in material coordinates. The results indicate that the **terminal metal velocity** (D_{metal}) is a function of the ratio of $\frac{m}{c}$ on the basis of an energy balance [80, 82]. For simple asymmetric configuration, a momentum balance is also required and must be solved simultaneously for simple asymmetric configuration [57, 59]. For some simple geometries filled with explosives, the metal velocity can be expressed as follows [80]:

- cylindrical tube:

$$\frac{D_{metal}}{\sqrt{2E_G}} = \left(\frac{m}{c} + \frac{1}{2}\right)^{-\frac{1}{2}};$$

(5.1)

- sphere:

$$\frac{D_{metal}}{\sqrt{2E_G}} = \left(\frac{m}{c} + \frac{3}{5}\right)^{-\frac{1}{2}};$$

(5.2)

- symmetrical sandwich:

$$\frac{D_{metal}}{\sqrt{2E_G}} = \left(\frac{m}{c} + \frac{1}{3}\right)^{-\frac{1}{2}}.$$

(5.3)

Thus, the measured values of D_{metal} with known geometry and $\frac{m}{c}$ can be used to correlate with $\sqrt{2E_G}$. Details of the application of the Gurney velocity for metal fragment projectiles from different charge geometries and the effect of air on the velocity of a piece of metal were demonstrated elsewhere [80].

5.2 Gurney energy and the cylinder expansion test

The **cylinder test** is a suitable experimental method to measure the effectiveness of an explosive. The radial expansion on detonation of a metallic cylinder (usually copper) which is filled with a high explosive can be observed using a streak camera or a laser method. It provides the most complete information to the warhead by viewing the expansion of the detonation products inside the cylinder tube. **Brisance** can be defined as the ratio between the potential of the explosive and the duration of the detonation. To estimate the explosive effects on a certain surrounding medium, it is preferable to use E_G instead of the brisance. If detonation is performed inside the tube of the cylinder test, the value of E_G consists of the kinetic energy of the expanding detonation and the kinetic energy of the displaced walls. The value of D_{metal} can be calculated by considering the detonation velocity taking into account the loss of kinetic energy of the detonation products and also heat loss according to [52]

$$D_{metal} = \frac{D_{det}}{2}\left(\frac{2m}{c}+1\right)^{-\frac{1}{2}}.$$ (5.4)

The cylinder test not only provides the value of E_G but it is also possible to calculate the detonation pressure and the heat of detonation if a more detailed consideration of the detonation products and the displacement of the wall material is undertaken.

5.2.1 Cylinder test measurements

The cylinder test provides a measure of the hydrodynamic performance of an explosive assuming a constant volume of high explosive. It consists of an explosive charge 25.4 mm in diameter and 0.31 m long in a tightly fitting copper tube with a wall thickness of 2.6 mm. The explosive charge is initiated at one end and the radial motion of the cylinder wall is measured at about 0.2 m from the initiated end using the streak camera technique. The transfer of kinetic energy to the copper wall in a fixed geometry leads to a simple way of expressing the performance of the explosive. **Detonation normal** or **head on** to the metal and **detonation tangential** or **sideways** to the metal are two extreme geometric arrangements for the transfer of explosive energy to adjacent metal [9]. Due to the effects of the equations of state of the detonation products, the effective explosive energy is frequently different for the two cases mentioned above. For both head-on and tangential detonation, the cylinder test provides a measure of the relative effective explosive energy. The radial wall velocities at 5–6 mm and

19 mm are indicative of explosive energies in head-on geometry and tangential geometry, respectively. Terminal wall velocities at breakup are about 7–10 % higher where approximately 50 % of the detonation energy is transferred to the cylinder wall.

5.2.2 Prediction methods of the cylinder test

Some thermochemical computer codes such as CHEETAH can be used to evaluate the cylinder test [133]. Several correlations have also been developed to estimate cylinder test outputs, which are described in the following sections.

5.2.2.1 Method based on the Kamlet–Jacobs decomposition products

Short et al. [134] used equations (1.5) and (1.10) to derive the wall velocity in the cylinder test of $C_aH_bN_cO_d$ explosives by a least squares fitting of the experimental data as follows:

$$V_{\text{cylinder wall}}$$
$$= 1.316\,\rho_0^{0.84}\,(n'_{\text{gas}}(\bar{M}_{\text{wgas}})^{0.5}(Q_{\text{det}}[H_2O(g)])^{0.5})^{0.54}(R - R_0)^{(0.212-0.065\rho_0)}, \qquad (5.5)$$

where $V_{\text{cylinder wall}}$ is the cylinder wall velocity in km/s, $R - R_0$ is actual radial expansion in mm and ρ_0 is the loading density in g/cm^3. For example, if the calculated data of HMX at ρ_0 = 1.894 g/cm^3 given in Section 3.4.1.1 is used, i. e. n'_{gas} = 0.03378 mol, $\bar{M}_{\text{wgas}}$ = 27.21 g/mol, and $Q_{\text{det}}[H_2O(g)]$ = 6.18 kJ/g, the value of $V_{\text{cylinder wall}}$ of HMX for $R - R_0$ = 6.0 and 19.0 mm can be estimated as follows:

$$R - R_0 = 6.0\,\text{mm}$$
$$\rightarrow V_{\text{cylinder wall}} = 1.316(1.894)^{0.84}(0.03378(27.21)^{0.5}(6.18)^{0.5})^{0.54}$$
$$\times (6.0)^{(0.212-0.065\times1.894)}$$
$$= 1.690\,\text{km/s},$$

$$R - R_0 = 19.0\,\text{mm}$$
$$\rightarrow V_{\text{cylinder wall}} = 1.316(1.894)^{0.84}(0.03378(27.21)^{0.5}(6.18)^{0.5})^{0.54}$$
$$\times (19.0)^{(0.212-0.065\times1.894)}$$
$$= 1.872\,\text{km/s}.$$

The measured values of the specific wall kinetic energies at 6 mm and 19 mm wall displacement for HMX are 1.410 and 1.745 MJ/kg, respectively and are characteristic of head-on and tangential detonation, respectively. The measured $V_{\text{cylinder wall}}$ can be easily obtained as follows:

$$V_{\text{cylinder wall}}(\text{head on} = 6.0\,\text{mm}) = (1.410 \times 2)^{0.5}$$
$$= 1.679\,\text{km/s},$$

$$V_{\text{cylinder wall}}(\text{tangential} = 19\,\text{mm}) = (1.745 \times 2)^{0.5}$$
$$= 1.868\,\text{km/s}.$$

Thus, the calculated $V_{\text{cylinder wall}}$ values are close to the experimental values. Short et al. [134] extended the detonation products of equation (1.10) for $C_aH_bN_cO_dF_e$ explosives by the following conditions:

(i) formation of HF: available hydrogen reacts firstly with fluorine to form HF;
(ii) formation of CF_4: remaining fluorine, if any, reacts with carbon to form CF_4;
(iii) formation of H_2O: any remaining hydrogen, (after (i)), reacts with oxygen to form H_2O;
(iv) formation of CO_2: any remaining oxygen reacts with carbon to form CO_2.

For several fluoro explosives, the results of tests have shown that equation (1.10), which assumes HF formation but not CF_4, provides relatively good predictions.

5.2.2.2 Improved method for prediction of cylinder wall velocity

It was shown that the decomposition pathways of equation (1.11) can be used to predict the cylinder wall velocity of $C_aH_bN_cO_dF_eCl_f$ explosives according to [135]

$$V_{\text{cylinder wall}}$$
$$= 1.262\rho_0^{0.84}\,(n'_{\text{gas}}(\bar{M}_{\text{w gas}})^{0.5}(Q_{\text{det}}[H_2O(g)])^{0.5})^{0.54}(R - R_0)^{(0.212-0.065\rho_0)}. \qquad (5.6)$$

For example, if the calculated data of HMX at $\rho_0 = 1.894\,\text{g/cm}^3$ given in Section 3.4.1.2 are used, i. e. $n'_{\text{gas}} = 0.04052\,\text{mol}$, $\bar{M}_{\text{w gas}} = 24.68\,\text{g/mol}$, and $Q_{\text{det}}[H_2O(g)] = 5.02\,\text{kJ/g}$, the value of $V_{\text{cylinder wall}}$ of HMX for $R - R_0 = 6.0$ and $19.0\,\text{mm}$ can be estimated as follows:

$$R - R_0 = 6.0\,\text{mm}$$
$$\rightarrow V_{\text{cylinder wall}} = 1.262(1.894)^{0.84}(0.04052(24.68)^{0.5}(5.02)^{0.5})^{0.54}$$
$$\times (6.0)^{(0.212-0.065\times1.894)}$$
$$= 1.646\,\text{km/s},$$

$$R - R_0 = 19.0\,\text{mm}$$
$$\rightarrow V_{\text{cylinder wall}} = 1.262(1.894)^{0.84}(0.04052(24.68)^{0.5}(5.02)^{0.5})^{0.54}$$
$$\times (19.0)^{(0.212-0.065\times1.894)}$$
$$= 1.824\,\text{km/s}.$$

The calculated values are close to the experimental values given in a previous section.

5.2.3 JWL equation of state

For metal acceleration, the Jones–Wilkins–Lee equation of state (JWL-EOS) can be used to accurately describe the pressure-volume-energy behavior of the detonation products of explosives; however, the values are valid only for large charges [9]:

$$
\begin{aligned}
P = A_{\text{JWL}}\left(1 - \frac{\omega}{(R_1 V_{\text{det}}/V_0)}\right)^{-(R_1 V_{\text{det}}/V_0)} \\
+ B_{\text{JWL}}\left(1 - \frac{\omega}{(R_2 V_{\text{det}}/V_0)}\right)^{-(R_2 V_{\text{det}}/V_0)} \\
+ \frac{\omega E}{(V_{\text{det}}/V_0)},
\end{aligned}
\tag{5.7}
$$

where A_{JWL} and B_{JWL} are linear coefficients in GPa; R_1, R_2, and ω are nonlinear coefficients; V_{det} and V_0 are the volumes of the detonation products and undetonated high explosive respectively; and P is in GPa and E is the detonation energy per unit volume in $(\text{GPa m}^3)/\text{m}^3$. This equation changes to the following equation at constant entropy or isentrope [136]:

$$
P_S = A_{\text{JWL}} e^{-(R_1 V/V_0)} + B_{\text{JWL}} e^{-(R_2 V/V_0)} + C_{\text{JWL}}(V/V_0)^{-\omega-1},
\tag{5.8}
$$

where C_{JWL} is also a linear coefficient in GPa. The coefficient ω is known as the Grüneisen coefficient or the second adiabatic coefficient, which is defined as [52, 137]

$$
\omega = -\left(\frac{\partial \ln T}{\partial \ln(V/V_0)}\right)_S.
\tag{5.9}
$$

The parameters A, B, C, R_1, R_2, and ω are determined by fitting the pressure-volume data to equation (5.6). These parameters have been determined for some explosives, which have been subjected to a rigorous comparison analysis by matching the results obtained from the equation with those obtained from experimental C–J conditions, calorimetric data, and expansion behavior – usually cylinder-test data [9].

5.3 Different methods for the prediction of the Gurney velocity

Suitable computer codes and appropriate equations of state can be used to predict the Gurney velocity. For example, Hardesty and Kennedy [82] have shown that Gurney velocities are reasonably well approximated by the TIGER computer code and the Jacobs–Cowperthwaite–Zwisler-3 equation of state (JCZ3-EOS) [138] as

$$
\sqrt{2E_{\text{G}}} = \left(\sqrt{2(U_0 - U_s)}\right)_{V/V_0=3}.
\tag{5.10}
$$

The JCZ3-EOS was formulated to allow it to be used generally for any explosive formulation of thermodynamic state properties of a mixture of product species for densities ranging from those at atmospheric pressure to those at the C–J state, as well as to permit a reliable estimation of the internal energy states during expansion [82]. Equation (5.10) shows that energy transfer between the detonation products and the driven metal is limited in many cases by rupturing of the metal rather than by side losses [82]. There are several empirical methods which can be used for the prediction of the Gurney velocity, which will be discussed in the following sections.

5.3.1 Using the Kamlet–Jacobs decomposition products

Hardesty and Kennedy (H–K) [82], as well as Kamlet and Finger (K–F) [139] have related the Gurney velocity to the decomposition products of Kamlet and Jacobs [5] (which are given in equation (1.10)) to predict the Gurney velocity as follows:

$$(\sqrt{2E_G})_{H-K} = 0.6 + 2.55(n'_{gas}\rho_0)^{0.5}(\bar{M}_{wgas}Q_{det}[H_2O(g)])^{0.25}, \tag{5.11}$$

$$(\sqrt{2E_G})_{K-F} = 3.49(n'_{gas})^{0.5}(\bar{M}_{wgas}Q_{det}[H_2O(g)])^{0.25}\rho_0^{0.4}, \tag{5.12}$$

where $(\sqrt{2E_G})_{H-K}$ and $(\sqrt{2E_G})_{K-F}$ are the Gurney velocities from the H–K [82] and K–F [139] methods. For example, if the calculated data of HMX at $\rho_0 = 1.89\,\text{g/cm}^3$ given in Section 3.4.1.1 are used, i.e. $n'_{gas} = 0.03378\,\text{mol}$, $\bar{M}_{wgas} = 27.21\,\text{g/mol}$ and $Q_{det}[H_2O(g)] = 6.18\,\text{kJ/g}$, the values of $(\sqrt{2E_G})_{H-K}$ and $(\sqrt{2E_G})_{K-F}$ are obtained as follows:

$$(\sqrt{2E_G})_{H-K} = 0.6 + 2.55(0.03378 \times 1.89)^{0.5}(27.21 \times 6.18)^{0.25}$$

$$= 2.92\,\text{km/s},$$

$$(\sqrt{2E_G})_{K-F} = 3.49(0.03378)^{0.5}(27.21 \times 6.18)^{0.25}(1.89)^{0.4}$$

$$= 2.98\,\text{km/s}.$$

The measured Gurney velocity of HMX is 2.97 km/s [9], which is closer to the value calculated for $(\sqrt{2E_G})_{H-K}$.

5.3.2 The use of elemental composition and the heat of formation

It was shown that the elemental composition of $C_aH_bN_cO_d$ explosives can be used to allow a reliable calculation of the Gurney velocity using the condensed or gas phase heats of formation of an explosive as follows [140]:

$$\sqrt{2E_G} = 0.227$$

$$+ \left(\frac{7.543a + 2.676b + 31.97c + 35.91d - 0.0468\Delta_fH^\theta(c)}{M_w}\right)\rho_0^{0.5}, \tag{5.13}$$

$$\sqrt{2E_G} = 0.220$$

$$+ \left(\frac{6.620a + 4.427b + 29.03c + 37.61d - 0.0122\Delta_f H^\theta(g)}{M_w} \right) \rho_0^{0.5}. \tag{5.14}$$

As is shown in the above correlations, the coefficients of $\Delta_f H^\theta(c)$ and $\Delta_f H^\theta(g)$ are small values compared to the coefficients for the elements, which have positive coefficients. Thus, the contribution of the four elements present in the unreacted explosive is far more important, in terms of influencing the Gurney velocity, than details of the bonding within the molecular structure. It was shown that the reliability of equations (5.13) and (5.14) are higher than both $(\sqrt{2E_G})_{H-K}$ and $(\sqrt{2E_G})_{K-F}$ [83]. For TNT, the measured $\Delta_f H^\theta(c)$ and the calculated $\Delta_f H^\theta(g)$ by B3LYP/6-31G* method are −67.01 kJ (Appendix) and 16.64 kJ/mol [91], respectively. Applying these data in equations (5.13) and (5.14) at a loading density of 1.63 g/cm^3 gives

$$\sqrt{2E_G} = 0.227$$

$$+ \left(\frac{7.543 \times 7 + 2.676 \times 5 + 31.97 \times 3 + 35.91 \times 6 - 0.0468 \times (-67.01)}{227.13} \right)(1.63)^{0.5}$$

$$= 2.37 \text{ km/s},$$

$$\sqrt{2E_G} = 0.220$$

$$+ \left(\frac{6.620 \times 7 + 4.427 \times 5 + 29.03 \times 3 + 37.61 \times 6 - 0.0122 \times 16.64}{227.13} \right)(1.63)^{0.5}$$

$$= 2.36 \text{ km/s}.$$

The predicted values of $(\sqrt{2E_G})_{H-K}$ and $(\sqrt{2E_G})_{K-F}$ are 2.43 and 2.38 km/s. Since the measured Gurney velocity is 2.37 km/s [9], the values calculated using equations (5.13) and (5.14) as well as $(\sqrt{2E_G})_{K-F}$ are close to the experimental values.

5.3.3 The use of elemental composition without using the heat of formation of an explosive

It was shown that the following equation is suitable for calculating the Gurney velocity of a $C_a H_b N_c O_d$ explosive [141] without requiring consideration of the heat of formation of the explosive:

$$\sqrt{2E_G} = 0.404 + 1.020\rho_0 - 0.021c + 0.184(b/d) + 0.303(d/a). \tag{5.15}$$

For example, as is shown in the Appendix, LX-14 with the composition 95.5/4.5 HMX/ Estane 5702-F1 has the formula $C_{1.52}H_{2.92}N_{2.59}O_{2.66}$ and $\Delta_f H^\theta(c) = 6.28$ kJ/mol. The use of equation (5.15) for LX-14 at a loading density of 1.68 g/cm^3 gives Gurney velocity as

$$\sqrt{2E_G} = 0.404 + 1.020 \times 1.68$$
$$- 0.021 \times 2.59 + 0.184(2.92/2.66) + 0.303(2.66/1.52)$$
$$= 2.80 \text{ km/s.}$$

The measured Gurney velocity for LX-14 is also 2.80 km/s [9], which is the same as the calculated value. The use of equations (5.11), (5.12), and (5.13) gives values of 2.74, 2.79, and 2.79 km/s, respectively, which shows the lower reliability of $(\sqrt{2E_G})_{H-K}$ with respect to other correlations. This correlation is the simplest method to calculate the Gurney velocity of explosives, but it may result in large deviations for those explosives in which $b = 0$.

5.4 Combined effects aluminized explosives

Thermobaric weapons create a large fireball and good blast performance [142]. They contain monopropellants or secondary explosives and additionally possess some elements such as Al [142]. The post-detonation reaction, the same as burning of Al, takes play with air after the explosion of the main charge of a **thermobaric/enhanced blast explosive** , which occurs and produces a huge "fireball" within a microsecond. **Enhanced blast explosives** attracted interest in recent years due to their effectiveness in buildings, bunkers, caves and other enclosed structures. Enhanced blast often decreases in the ability of the explosive to push the metal or Gurney velocity, which is due to the delayed oxidation of aluminum until later in the fireball expansion, generally well above 10–50 volume expansions. Traditional high-energy aluminized explosives commonly incorporate some percentages of aluminum. They can produce high blast energies during their detonation but they are normally characterized with the reduced metal pushing capability due to the relatively late-time aluminum reaction.

For aluminized explosives with an energetic binder system and small aluminum particles, the term **"combined effects"** has been recently developed because the formulations show both the enhanced blast and impressive Gurney velocities at volume expansions $V/V_0 < 10$ [143]. While the systems of "combined effects" can be optimized for metal pushing according to particle size, binder content, and other parameters, the exact mechanism of how aluminum reacts early in the detonation event is still under inspection. It has been previously hypothesized that higher oxygen balance materials and a proper choice of the aluminum particle size lead to the early aluminum reaction [144–146]. Upon the early reaction of aluminum, a shift in the gaseous reaction products also comes to higher enthalpy species such as CO and H_2, which leads to a further augmentation of blast. The enhanced metal-pushing capability is due to the earlier exothermic conversion of Al to aluminum oxide as compared to the conventional blast explosives. Thus, a composite aluminized explosive under this situation can provide both the mechanical energy for fragmentation or "metal pushing" and blast or structural targets here.

The oxygen balance may be a contributing factor to the efficiency of afterburning and blast. Since oxygen is diffusion-limited in the detonation reaction zone, the oxygen balance may not necessarily contribute to the immediate combustion of fuels. This situation can occur due to the diffusion time limitations at the molecular level during the detonation event [147]. It has been indicated that energetic binders can provide the early aluminum reaction [148]. For the presence of excess oxygen close to the aluminum particles, such as with energetic binders, there is an increasing likelihood that the oxygen will promote the early aluminum reaction. It was found that very little aluminum reacted when the detonation was carried out in an inert atmosphere for detonation calorimetry experiments utilizing HTPB at various oxygen levels [147] as it worked by the time resolved emission spectroscopy [147]. Anderson et al. [146, 147] used a mixture **design of experiments (DOE)** to determine what compositional variables promote or preclude the anaerobic reaction of Al in enhanced blast explosives and combined effects explosives. They investigated the effect of binder type, binder content, explosive content, Al content and Al particle size. They found that there is only a general relationship between the %Al reaction and oxygen balance. It should be mentioned that the traditional C–J detonation theory with completely reacted aluminum does not explain the observed detonation states achieved by these combined-effects explosives [149–151]. Due to inability of Al to participate in the reaction zone, it behaves as chemically inert Al. Thus, the detonation velocity and detonation pressure are higher than they would have been if the Al had been allowed to participate in the reaction zone. It was demonstrated that the eigenvalue detonation theory explains the observed behavior for combined-effects aluminized explosives in which both high metal pushing capability and high blast are achieved [150].

5.5 Assessment of the Gurney velocity of aluminized and nonaluminized explosives

Different methods in the previous section can estimate the Gurney velocity of CHNO-based explosives. Since some of them can give relatively large deviations from the experiment when applied to CHNOAl-based explosives, there are some correlations for prediction of the Gurney velocity of aluminized and nonaluminized explosives. These models are more complex and require more input data than that discussed here.

5.5.1 Using the BKW thermochemical code

A formula for predicting the Gurney velocity of pure, mixed, and aluminized high explosives at any initial density was proposed in [152]. This formula requires the calculated detonation pressure and adiabatic exponent by the BKW code of Mader [12] using the RDX set of parameters as [152]:

$$\sqrt{2E_G} = (\ln P_{det} - \gamma)_{BKW}, \tag{5.16}$$

where P_{det} is in kbar. For example, the computed P_{det} and y using the output of the BKW code [12] for HBX-3 at $\rho_0 = 1.81\,g/cm^3$ are 195 kbar and 3.37, respectively. The use of these values in equation (5.16) gives:

$$\sqrt{2E_G} = (\ln 195 - 3.37)_{BKW} = 1.90\,km/s.$$

The measured value of the Gurney velocity is 1.98 km/s [152].

5.5.2 Detonation pressure and the specific impulse

The specific impulse shows a measure of how effectively a rocket uses propellant or a jet engine uses fuel [153, 154]. It is defined as the total impulse (or change in momentum) delivered per unit of propellant consumed. It was indicated that the detonation pressure and the specific impulse can be used to estimate the Gurney velocity as [155]:

$$\sqrt{2E_G} = 0.02415\left(\frac{P_{det}}{I_{SP}\rho_0}\right) + 0.97076, \tag{5.17}$$

where P_{det} is in kbar, ρ_0 is in g/cm^3 and I_{SP} is the specific impulse of the explosive used as a monopropellant in N.s/kg. Different methods given in Chapter 4 or computer codes may be used for calculation of the detonation pressure. The simplest available method of the following form can also be used to estimate the specific impulse of $C_aH_bN_cO_d$ explosives as [140]:

$$I_{SP} = 2.4205 - 0.0740a - 0.0036b + 0.0237c + 0.0400d$$
$$- 0.1001n_{NH_x} + 0.1466(n_{Ar} - 1), \tag{5.18}$$

where I_{SP} is in N.s/kg, n_{NH_x} is the number of $-NH_2$ and $-NH$ groups and n_{Ar} is the number of aromatic rings in aromatic pure explosives. It should be mentioned that the value of the last term in equation (5.18) should be taken zero for nonaromatic pure explosives, i. e. $0.1466(n_{Ar} - 1) = 0.0$. For example, the calculated detonation pressure for HMX in Section 4.3.1.2 at $\rho_0 = 1.89\,g/cm^3$ is 384.3 kbar. Equation (5.18) gives the specific impulse for HMX with formula $C_4H_8N_8O_8$ as:

$$I_{SP} = 2.4205 - 0.0740 \times 4 - 0.0036 \times 8 + 0.0237 \times 8 + 0.0400 \times 8 - 0.1001 \times 0 + 0$$
$$= 2.605\,N.g/kg.$$

The use of the calculated detonation pressure and the specific impulse in equation (5.17) provides:

$$\sqrt{2E_G} = 0.02415\left(\frac{384.3}{2.605 \times 1.89}\right) + 0.97076 = 2.856 \text{ km/s}.$$

The calculated value is close to the experimental data, i. e. 2.970 km/s [9].

5.5.3 The Gurney velocity based on the heat of combustion

It was shown that the following correlation based on four variables can be used to estimate the Gurney velocity as follows [143]:

$$\sqrt{2E_G} = 2.261 + 0.5941\rho_0 + 0.0448\Delta H_c^\theta - 0.1394\tau + 0.30\varphi, \tag{5.19}$$

where ρ_0 is the loading density in g/cm^3, ΔH_c^θ is the heat of combustion in kJ/mol, τ and φ are two dummy variables, which can be found from chemical formula and behavior of composite explosives. For calculation of the heat of combustion, equation (1.6) can be improved to consider the presence of Al as follows:

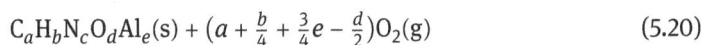

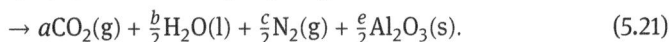

$$C_aH_bN_cO_dAl_e(s) + (a + \tfrac{b}{4} + \tfrac{3}{4}e - \tfrac{d}{2})O_2(g) \tag{5.20}$$
$$\rightarrow aCO_2(g) + \tfrac{b}{2}H_2O(l) + \tfrac{c}{2}N_2(g) + \tfrac{e}{2}Al_2O_3(s). \tag{5.21}$$

For the presence of further chemical elements such as calcium (Ca) or phosphorus (P) in little amounts in some explosives, it is better to consider their oxide forms including CaO and P_4O_{10} for calculation of the heat of combustion. Hydrogen halides such as hydrogen fluoride (HF) and hydrogen chloride (HCl) should be considered as combustion products for the existence of corresponding halides, i. e. fluorine (F) and chlorine (Cl), in chemical composition of explosive. Then, the heat of combustion is calculated using the heat of formation of the products and explosive according to equations (5.21) and (1.5) as:

$$\Delta H_c^\theta = \frac{\sum n_j \Delta_f H^\theta (\text{detonation product})_j - \Delta_f H^\theta (\text{explosive})}{\text{formula weight of explosive}}. \tag{5.22}$$

The parameter τ shows the total weight fraction of nitroaromatic explosives present in nonaluminized composite explosive. For those aluminized explosives containing nitroaromatic, the value of τ equals the weight fraction of the aluminum plus that of the nitroaromatic. The parameter φ has the value 1.0 if the studied composition is a combined effects explosive otherwise it takes the value of zero. In contrast to the detonation velocity and pressure, the predicted Gurney velocities of the combined effects explosives such PAX-29n and PAX-30 should be taken higher than for conventional explosives because further factors can influence the Gurney velocity. For example, PAX-30 and Octol 75/25 have the same detonation velocity and pressure but the Gurney velocity of PAX-30 is 8.5 % greater than that of Octol 75/25. The value $\varphi = 1.0$ should be taken if all of the following conditions are satisfied:

(a) the existence of an "energetic" binder such as cellulose acetate butyrate plasticized with BDNPA/F;
(b) the use of aluminum with small particle sizes (<50 mm); and
(c) the existence of large amounts of a major explosive material (>75 wt%).

Thus, if any one of three conditions is not met, then the explosive is classified as a traditional explosive where the value of φ should be equal to zero (e. g., PAX-3a, HTA-3, HBX-1). For example, consider PAX-29n with composition 77/4.8/15/3.2 Cl-20/BD-NPA/F/submicron-Al/CAB and chemical formula $(C_{1.313}H_{1.474}N_{2.169}O_{2.336}Al_{0.556})$, $\Delta_f H^\theta = 43.14$ kJ/mol and $\rho_0 = 2.010$ g/cm^3. The use of equations (5.19) and (5.21) gives the heat of combustion as:

$$\Delta H_c^\theta = \frac{(1.313)(\Delta_f H^\theta[CO_2(g)]) + (1.474/2)(\Delta_f H^\theta[H_2O(l)]) + (0.556/2)(\Delta_f H^\theta[Al_2O_3(s)])}{100 \text{ g/mol}}$$

$$= \frac{(1.313)(-393.5 \text{ kJ/mol}) + (0.737)(-285.8 \text{ kJ/mol}) + (0.278)(-1676 \text{ kJ/mol})}{100 \text{ g/mol}}$$

$$= -12.36 \text{ kJ/g}.$$

Since $\tau = 0$ and $\varphi = 1.0$, equation (5.19) gives:

$$\sqrt{2E_G} = 2.261 + 0.5941(2.010) + 0.0448(-12.36) - 0.1394(0) + 0.30(1) = 3.20 \text{ km/s}.$$

The calculated Gurney velocity is close to the reported value, i. e. 3.16 km/s [9].

Summary

The specific energy according to Gurney's model and the related Gurney velocity are better values for the ballistic characterization of an explosive than the detonation properties are, because they allow calculation of the velocities and impulses imparted to driven materials. Several empirical methods have been reviewed for the calculation of the cylinder wall velocity, as well as the velocity of explosively-driven metal over a range of geometries and loading factors, which are of more practical importance to the explosive user. The correlations which have been introduced can be applied to pure explosives, as well as to solid explosive mixtures.

Also, equations (5.16), (5.17) and (5.19) have been developed for prediction of the Gurney velocity of CHNO explosives, as well as of CHNOAl-based ones.

Questions and problems

i The necessary information for some problems are given in the Appendix.

(1) Calculate the cylinder wall velocity of LX-09-0 with the composition: HMX (93 %), DNPA-F (4.6 %), FEFO (2.4 %), $C_{1.43}H_{2.74}N_{2.59}O_{2.72}F_{0.02}$; and solid phase heat of formation: 7.61 kJ/mol at loading density 1.836 g/cm^3 for $R - R_0 = 6.0$ and 19.0 mm using equation (5.6).

(2) For COMP A-3 calculate the Gurney velocity at loading density 1.59 g/cm^3 by
   (a) equations (5.11), (5.12), and (5.13);
   (b) equation (5.14), if the gas phase heat of formation of COMP A-3 is 142.3 kJ/mol.

(3) Use equation (5.15) to calculate the Gurney velocity of MEN-II at loading density 1.02 g/cm^3.

(4) The computed P_{det} and y using the output of the BKW code [12] for Tritonal (80/20) at $\rho_0 = 1.72$ g/cm^3 are 196 kbar and 2.92, respectively. Calculate the Gurney velocity of Tritonal (80/20).

(5) Consider Minol II with composition 40/40/20 AN/TNT/Al ($C_{1.233}H_{2.880}N_{1.528}O_{2.556}$ $Al_{0.741}$), $\Delta_f H^\theta = -194.23$ kJ/mol and $\rho_0 = 1.680$ g/cm^3. Calculate the Gurney velocity of Minol II.

i For answers and solutions, please see p. 120

6 Power (strength)

The detonation process for an explosive liberates energy, which can be used for performing mechanical work. The energy of an explosive is defined as the total work, i. e. the **maximum work** or **explosive potential** that may be performed by the hot gaseous products. The internal energy of the detonation products is completely transformed into mechanical work in this situation. Thus, the performance potential of an explosive can be described by three parameters:

(i)   the volume of gas liberated per unit weight,
(ii) the energy (heat) evolved in the process,
(iii) the propagation rate (velocity) of the explosive.

The performance of an explosive is specified in terms of the velocity of the detonation, detonation pressure, and **power** [1]. In general, a high gas yield and high heat of detonation are the two important factors necessary in order to obtain a high detonation performance [1]. If detonation occurs in the air, the mechanical work is nearly equal to the heat of detonation. For the detonation of an explosive in a borehole, the relevant parameter is its **power** or **strength** (also called the **blasting capacity** or **energy of detonation products**), which is a measure of the ability of an explosive to do useful work [52]. In rock blasting, for example, part of the energy of the products is expended in heating the rock and the other part remains in the products as thermal energy. The mechanical work in this case is always lower than the heat of detonation and accounts for 70–80 % of its value [1]. Thus, the assessment of the performance of an explosive based on its power is not so much dependent on a high detonation rate as it is on a high gas yield and large amount of energy evolved. As mentioned in Chapter 5, **brisance** is required for a strong disintegration effect in the vicinity of the detonation, because the most important parameters are the detonation rate and the loading density (compactness) of the explosive. A number of conventional tests and computational methods exist for determining the comparative power and brisance of different explosives, which will be described in this chapter.

6.1 Different methods for measuring the power and brisance of an explosive

Although explosive power and brisance can be determined directly from field measurements, laboratory methods are preferred. There are several experimental methods which can be used for the determination of the power and brisance of an explosive. In these methods, the work capacity is not expressed in work units, but usually in the change in a desired parameter instead, such as the increase in volume after explosion

https://doi.org/10.1515/9783110677652-006

within the lead block. The usual methods for determining the power include
(a) the lead block test,
(b) the ballistic mortar test,
(c) underwater detonation.

The common methods used to assess the brisance of an explosive are
(a) the sand crush test and
(b) the plate dent test [1, 52].

Among the different tests for predicting power, the Trauzl lead block and the ballistic mortar tests are the two most well known for assessing the energy released by detonating explosives. The Trauzl lead block test is one of the conventional laboratory tests and measures the expansion caused by an explosion contained within a block of lead or aluminum. As the amount of explosive energy released increases, a larger amount of expansion can be expected. The Trauzl test consists of a standard cast cylindrical lead block of 200 mm height and 200 mm diameter that contains 10 g of the explosive material and a detonator. The cylindrical lead block has an axial recess of 25 mm diameter and 125 mm depth that is stemmed with quartz sand. After firing the shot, the volume increase of the cavity is recorded (Figure 6.1).

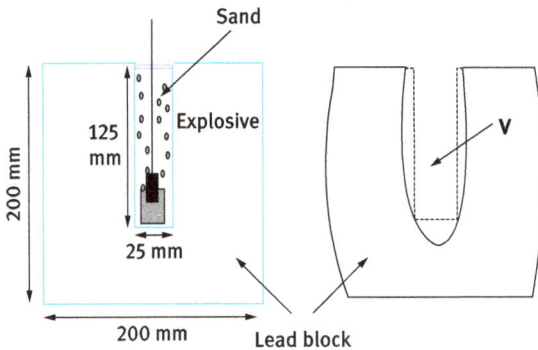

Figure 6.1: The Trauzl test of an explosive: (a) before the test and (b) after the test.

The ballistic mortar test provides a relative measure of the power of explosives by comparison usually with TNT or blasting gelatine as the standard explosive. In this test, a heavy steel mortar is attached to a pendulum. About 10 g of the explosive charge is initiated in a massive mortar enclosed by a projectile. The mortar is swung from its position when the projectile is ejected out of the mortar. The maximum swing of the mortar is recorded to determine the power of a desired explosive (Figure 6.2).

For commercial blasting and many other applications for which the power is important, the loading density is near unity. The Trauzl lead block method is the most

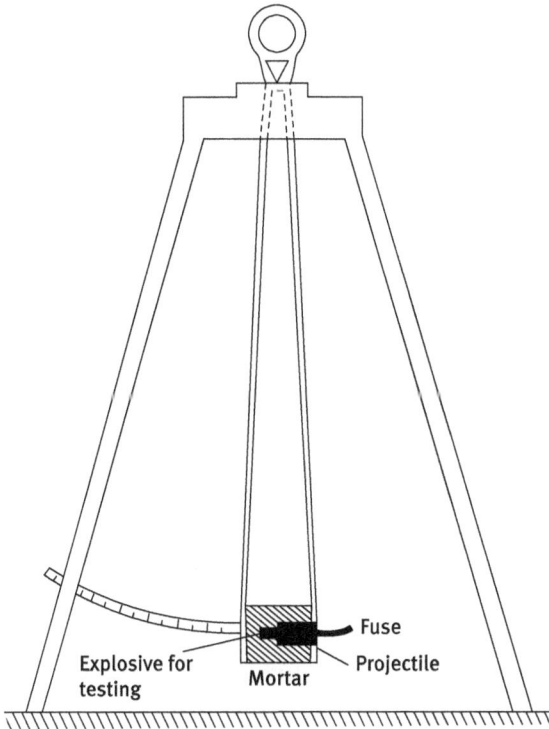

Figure 6.2: The ballistic mortar test.

widely used method for determining the power of high explosives because it provides data at loading densities which are comparable to those used in practice [1].

For measuring the brisance of an explosive, the sand test or sand crushing test is a suitable method [156]. It is based on determining the amount of standard sand which is crushed by a standard mass of explosive. Since this method is simple and can be applied for different pure, mixed and aluminized explosives, it is a convenient method for determining the brisance of energetic compounds in comparison with methods [156].

The **plate dent test** provides an adequate tool for obtaining the brisance by carefully examining the capability of a detonating explosive to impart a dent or a depression on a steel plate or any other suitable metal. Thus, an explosive with a more brisant can provide a deeper dent. The plate dent test permits the calculation of the detonation pressure using an empirical correlation. Figure 6.3 shows the plate dent test which contains an unconfined cylinder of 41.3 mm diameter and 203 mm length that rests vertically on a square plate 152.4 × 152.4 mm and 50.8 mm thickness made of 1018 cold-rolled steel with a Rockwell hardness of B-74 to B-76 [157]. Several other steel plates support the test plate itself to eliminate spalling from the rear surface [158]. A booster

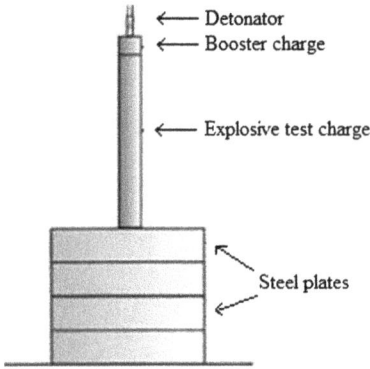

Figure 6.3: The plate dent test.

charge is placed on the top of the cylinder. When the whole assembly is fired with a detonator, the dent produced on the steel plate is measured using a depth micrometer.

6.2 Different methods for the prediction of power

The explosive power of an explosive is obtained by multiplying the heat of detonation and the volume of gaseous products ($V_{\text{exp gas}}$) according to [3, 7]:

$$\text{explosive power} = Q_{\text{expl}} V_{\text{exp gas}}. \tag{6.1}$$

Thermochemical computer codes such as EXPLO5 [10] can be used to obtain Q_{expl} and $V_{\text{exp gas}}$ for the calculation of the power. It is also possible to use approximate values of Q_{det} and the number of moles of the assumed detonation products. The volume of the gaseous products liberated during an explosion by a mixed explosive composition at "STP" (standard temperature and pressure) can be calculated from the assumed gaseous detonation products, i. e. 1 mol of gas occupies 22.4 l at a pressure of 1 atm and a temperature of 0 °C = 273.15 K.

As was shown in Chapter 1, different empirical methods have been developed for the prediction of Q_{det}. The volume of explosion gases is usually expressed in liters per kg of explosive material and is the volume of the gases (fumes) formed by the explosive reaction [1]. It is calculated from the chemical composition of the explosive by computer codes such as EXPLO5 [10] through the calculation of the moles of gaseous products. The Bichel bomb [1] can be used to determine experimentally the volume of gaseous products. In practice, the composition of the products and the volume of gaseous products are determined at the point of "frozen" chemical equilibrium, i. e. after fast cooling of the explosion products preferably to the STP conditions [52].

6.2.1 A simple correlation for the prediction of the volume of explosion gases of energetic compounds

It was shown that the following correlation gives the volume of explosion gases of energetic compounds with general formula $C_a H_b N_c O_d$ [159]:

$$V_{exp\ gas} = 878.2 - 126.0(a/d) + 111.7(b/d) - 176.7V_{corr}, \tag{6.2}$$

where $V_{exp\ gas}$ is the volume of explosion gases in l/kg, and the parameter V_{corr} is the correcting function. The value of V_{corr} is 1.0 for $C_a H_b N_c O_d$ explosives which fulfill both of the following conditions:
(i) a is not zero and
(ii) $d - 2a - b/2 \geq 0$.

The reliability of this new method is also higher than the outputs of complex computer codes using two of the best available equations of states, i. e. the BKW-EOS and the Jacobs–Cowperthwaite–Zwisler (JCZ-EOS) [11, 160]. For 69 energetic materials it was shown that the root-mean-square error (RMSE) of the new correlation shows relatively good agreement with experimental values, and is equal to 55 l/kg, as compared to the values of 86 and 116 l/kg, which were obtained using the BKW-EOS and JCZ3-EOS, respectively [159].

For example, the calculated $V_{exp\ gas}$ of [2-nitro-3-(nitrooxy)-2-[(nitrooxy)methyl] propyl]-nitrate (NIBTN) with the empirical formula $C_4 H_6 N_4 O_{11}$ is given as follows:

$$V_{exp\ gas} = 878.2 - 126.0(4/11) + 111.7(6/11) - 176.7(1)$$
$$= 717\ l/kg.$$

The value of $V_{exp\ gas}$ is 1.0 because NIBTN follows the two conditions, i. e.
(i) $a = 4$ and
(ii) $d - 2a - b/2 = 11 - 2(4) - (6/2) = 0$.

The calculated $V_{exp\ gas}$ is close to the measured value (705 l/kg) [1].

6.2.2 Power index

Since there are many different approaches for calculation of Q_{det} and $V_{exp\ gas}$, it is important to compare the power of a particular explosive with that of a standard explosive. For this purpose, the power index is defined as follows:

$$\text{power index}[H_2O(g)] = \frac{Q_{det}[H_2O(g)] \times V_{exp\ gas}}{(Q_{det}[H_2O(g)])_{picric\ acid} \times (V_{exp\ gas})_{picric\ acid}}, \tag{6.3}$$

$$\text{power index}[H_2O(l)] = \frac{Q_{det}[H_2O(l)] \times V_{exp\ gas}}{(Q_{det}[H_2O(l)])_{picric\ acid} \times (V_{exp\ gas})_{picric\ acid}}, \tag{6.4}$$

where power index[$H_2O(g)$] and power index[$H_2O(l)$] are the power indexes of an explosive if the water present in the detonation products is in the gaseous and liquid states, respectively. The use of experimental values of $Q_{det}[H_2O(g)] = 3350$ kJ/kg, $Q_{det}[H_2O(l)] = 3437$ kJ/kg, and $V_{exp\,gas} = 826$ l/kg [1] gives

$$\text{power index}[H_2O(g)] = \frac{Q_{det}[H_2O(g)] \times V_{exp\,gas}}{3350\,\frac{kJ}{kg} \times 826\,\frac{l}{kg}}$$

$$= \frac{Q_{det}[H_2O(g)] \times V_{exp\,gas}}{2.767 \times 10^6\,\frac{kJl}{(kg)^2}}, \tag{6.5}$$

$$\text{power index}[H_2O(l)] = \frac{Q_{det}[H_2O(l)] \times V_{exp\,gas}}{3437\,\frac{kJ}{kg} \times 826\,\frac{l}{kg}}$$

$$= \frac{Q_{det}[H_2O(l)] \times V_{exp\,gas}}{2.839 \times 10^6\,\frac{kJl}{(kg)^2}}. \tag{6.6}$$

Since different empirical approaches exist for $Q_{det}[H_2O(g)]$ and $Q_{det}[H_2O(l)]$, as was shown in Chapter 1, the use of these equations and equation (6.2) can be used to predict the power of an explosive.

6.2.3 Simple correlations for the prediction of power on the basis of the Trauzl lead block and the ballistic mortar tests

There are several empirical methods which can be used for the prediction of the power of pure and composite explosives based on the Trauzl lead block and the ballistic mortar tests. The relative power determined by these methods is usually compared with a desired energetic compound such as TNT. For the Trauzl lead block test, the relative power of an energetic compound with respect to TNT ($\%f_{Trauzl,TNT}$) is given by

$$\%f_{Trauzl,TNT} = \frac{\Delta V_{Trauzl}(\text{energetic compound})}{\Delta V_{Trauzl}(TNT)} \times 100, \tag{6.7}$$

in which ΔV_{Trauzl}(energetic compound) and ΔV_{Trauzl}(TNT) are the volume of expansion for the explosive and TNT, respectively. If the average experimental value of $\Delta V_{Trauzl}(TNT) = 295$ cm^3 [159] is used, equation (6.7) changes to

$$\%f_{Trauzl,TNT} = \frac{\Delta V_{Trauzl}(\text{energetic compound})}{295\,\text{cm}^3} \times 100. \tag{6.8}$$

The value $\%f_{Trauzl,TNT}$ can be used to qualitatively characterize the power of an energetic compound, if the error associated with it remains less than the error associated with ΔV_{Trauzl}. Therefore, it can be used for the quantitative comparison of the effect of explosions for different energetic compounds. For the ballistic mortar test, the maximum swing of the mortar of a particular energetic compound relative to TNT is also used and denoted by $\%f_{ballistic\,mortar,TNT}$ here.

6.2.3.1 The Trauzl lead block test

Several methods have been developed for the prediction of $\%f_{Trauzl,TNT}$ on the basis of different variables.

Specific impulse and heat of detonation

The specific impulse is used as a key measure of propellant performance. It can be defined as the integral of the thrust per unit mass of compound over the time of combustion. Due to the release of gaseous products during combustion, an energetic compound develops thrust. Molecular structures of energetic compounds can be used to predict their specific impulse [163, 164]. It was shown that the detonation pressure and detonation velocity can be related to the theoretical specific impulse [92, 93]. It is an appropriate way for estimating the performance of a wide variety of ideal and less ideal explosives. It was found that $\%f_{Trauzl,TNT}$ can be related to the specific impulse (I_{SP}) and the heat of detonation for a wide range of energetic compounds by [165]:

$$\%f_{Trauzl,TNT} = -97.25 + 18.87Q_{det}[H_2O(l)] + 25.69I_{SP}, \tag{6.9}$$

where $Q_{det}[H_2O(l)]$ and I_{SP} are in kJ/g and N s/g, respectively. Different computer codes, such as the standard Naval Weapons Center propellant performance code [166], NASA program computations [167], and ISPBKW [12], or empirical methods [161, 162], can be used to calculate the specific impulse of energetic compounds. For example, the calculated specific impulse for urea nitrate ($CH_5N_3O_4$) is 2.178 N s/g. As is shown in Table 1.2, the calculated $Q_{det}[H_2O(l)]$ for urea nitrate using equations (1.19), (1.20), and (1.22) is 3.484, 3.69, and 3.789 kJ/g, respectively. The use of these values in equation (6.9) gives

$$\%f_{Trauzl,TNT} = -97.25 + 18.87(3.484) + 59.59(2.178)$$
$$= 98.28,$$

$$\%f_{Trauzl,TNT} = -97.25 + 18.87(3.69) + 59.59(2.178)$$
$$= 102.2,$$

$$\%f_{Trauzl,TNT} = -97.25 + 18.87(3.789) + 59.59(2.178)$$
$$= 104.0.$$

These values are close to the measured value $\%f_{Trauzl,TNT}$ of urea nitrate, which is 98.31 [1].

Heat of detonation

For an energetic compound with general formula $C_aH_bN_cO_d$, it was found that $\%f_{Trauzl,TNT}$ can be calculated by [168]

$$\%f_{Trauzl,TNT} = -45.88(a/d) + 26.23Q_{det}[H_2O(l)], \tag{6.10}$$

where $Q_{det}[H_2O(l)]$ is in kJ/g. For example, as was shown in Section 1.2.1, $Q_{det}[H_2O(l)]$ is 6.77 kJ/g for HMX, which gives the $\%f_{Trauzl,TNT}$ value as

$$\%f_{Trauzl,TNT} = -45.88(4/8) + 26.23(6.77)$$
$$= 155.$$

The calculated value of $\%f_{Trauzl,TNT}$ is close to the measured value, whereby $\%f_{Trauzl,TNT}$ of HMX is in the range 145–163 [169].

The condensed and gas phase heats of formation
It was shown that the condensed and gas phase heats of formation of an energetic compound with general formula $C_aH_bN_cO_d$ can be directly used to calculate $\%f_{Trauzl,TNT}$ as follows [170]:

$$\%f_{Trauzl,TNT} = 471.2 + \frac{-8095a - 8993c + 38.71\Delta_fH^\theta(g)}{\text{formula weight of explosive}}, \tag{6.11}$$

$$\%f_{Trauzl,TNT} = 373.2 + \frac{-6525a - 5059c + 21.74\Delta_fH^\theta(c)}{\text{formula weight of explosive}}. \tag{6.12}$$

For example, as is shown in Section 1.3.1.1 and the Appendix, the values of $\Delta_fH^\theta(g)$ and $\Delta_fH^\theta(c)$ for HNS ($C_{14}H_6N_6O_{12}$) are 150 and 78.24 kJ/mol, respectively. The use of these values in equations (6.11) and (6.12) gives

$$\%f_{Trauzl,TNT} = 471.2 + \frac{-8095(14) - 8993(6) + 38.71(150)}{450.23}$$
$$= 112,$$

$$\%f_{Trauzl,TNT} = 373.2 + \frac{-6525(14) - 5059(6) + 21.74(78.24)}{450.23}$$
$$= 107.$$

These values are close to the measured value of HNS, i. e. $\%f_{Trauzl,TNT} = 102$ [1].

Molecular structure of an energetic compound
For energetic compounds with general formula $C_aH_bN_cO_d$, it was shown that their molecular structures can be used for the prediction of $\%f_{Trauzl,TNT}$ as follows [171]:

$$\%f_{Trauzl,TNT} = 196.2 - 59.46(a/d) - 30.13(b/d)$$
$$+ 47.56f^+_{Trauzl} - 41.49f^-_{Trauzl}, \tag{6.13}$$

where f^+_{Trauzl} and f^-_{Trauzl} are two correcting functions for the adjustment of the underestimated and the overestimated values of $\%f_{Trauzl,TNT}$ obtained on the basis of the elemental composition. Table 6.1 shows the values of f^+_{Trauzl} and f^-_{Trauzl} in different pure energetic compounds.

Table 6.1: Values of f_{Trauzl}^+ and f_{Trauzl}^-.

Molecular moieties	f_{Trauzl}^+	f_{Trauzl}^-	Condition
$R-(ONO_2)_x$, $x = 1, 2$	1.0	–	Without the other functional groups except $-C-NO_2$
$R-(ONO_2)_x$, $x \geq 3$	0.5	–	
$R-(NNO_2)_x$, $x = 1, 2, \ldots$	0.5	–	
(structure) $-(NO_2)_x$, $x \leq 2$	0.8	–	–
(structure) H_2N–$C(O)$–N^-H	–	1.0	–
phenyl-$(OH)_x$ or phenyl-$(ONH_4)_x$	–	0.5x	–
phenyl-$(NH_2)_x$ or phenyl-$(NHR)_x$	–	0.4x	–
phenyl-$(OR)_x$	–	0.2x	–
phenyl-$(COOH)_x$	–	0.9x	–

The following equation can be used to obtain acceptable results for mixtures of energetic compounds:

$$\%f_{Trauzl,TNT} = \sum_j x_j (\%f_{Trauzl,TNT})_j, \tag{6.14}$$

where x_j is the mole fraction of the j-th component in the energetic compound mixture. For example, the value of $\%f_{Trauzl,TNT}$ for pentolite-50/50 containing 50 % PETN ($C_5H_8N_4O_{12}$) and 50 % TNT is calculated from equations (6.13) and (6.14) as

$$(\%f_{Trauzl,TNT})_{PETN} = 196.2 - 59.46(5/12) - 30.13(8/12) + 47.56(0.5) - 41.49(0)$$

$$= 174,$$

$$\%f_{Trauzl,TNT} = x_{TNT}(\%f_{Trauzl,TNT})_{TNT} + x_{PETN}(\%f_{Trauzl,TNT})_{PETN}$$

$$= \left(\frac{\frac{50}{227.13}}{\frac{50}{227.13} + \frac{50}{316.14}} \right)(100) + \left(\frac{\frac{50}{316.14}}{\frac{50}{227.13} + \frac{50}{316.14}} \right)(174)$$

$$= (0.582)(100) + (0.418)(174)$$

$$= 131.$$

The measured $\%f_{Trauzl,TNT}$ for pentolite-50/50 is 122 [172].

6.2.3.2 The ballistic mortar test

To assess the power of energetic compounds with the general formula $C_aH_bN_cO_d$ through the ballistic mortar test, the following correlation can be used [173]:

$$\%f_{ballistic\ mortar,TNT} = 113 - 5.16a + 2.79c + 3.61d - 46.18f_{ballistic\ mortar}^-. \tag{6.15}$$

The parameter $f_{\text{ballistic mortar}}^-$ can be used to correct overestimated values based on the elemental composition as follows.

(i) Dinitrosubstituted benzene or those energetic compounds with the condition $d - (a + \frac{b}{2}) \geq 8$: the value of $f_{\text{ballistic mortar}}^-$ is 0.7.

(ii) Specific energetic compounds: For energetic compounds complying with one of the following conditions (1) $a = 0$, (2) $b = 0$, (3) the presence of the molecular fragment $-NH-CO-NH-$: the value of $f_{\text{ballistic mortar}}^-$ equals 1.0.

The following equation can be used to obtain acceptable results for mixtures of energetic compounds:

$$\%f_{\text{ballistic mortar,TNT}} = \sum_j x_j (\%f_{\text{ballistic mortar,TNT}})_j. \tag{6.16}$$

For example, the value of $\%f_{\text{ballistic mortar,TNT}}$ for AMATOL 80/20 containing 80 % NH_4NO_3 (AN) and 50 % TNT is calculated from equations (6.15) and (6.16) as

$$(\%f_{\text{ballistic mortar,TNT}})_{\text{AN}} = 113 - 5.16(0) + 2.79(2) + 3.61(3) - 46.18(0)$$

$$= 129,$$

$$\%f_{\text{ballistic mortar,TNT}} = x_{\text{TNT}}(\%f_{\text{ballistic mortar,TNT}})_{\text{TNT}} + x_{\text{AN}}(\%f_{\text{ballistic mortar,TNT}})_{\text{AN}}$$

$$= \left(\frac{\frac{50}{227.13}}{\frac{50}{227.13} + \frac{50}{80.04}} \right)(100) + \left(\frac{\frac{50}{80.04}}{\frac{50}{227.13} + \frac{50}{80.04}} \right)(129)$$

$$= (0.261)(100) + (0.739)(129)$$

$$= 121.$$

The experimental value of $\%f_{\text{ballistic mortar,TNT}}$ for AMATOL 80/20 is 130 [174].

6.3 Prediction of brisance

Upon detonation of an explosive, high pressure is created in its shock wave that will shatter rather than displace any object in its path, while subsequent expansion of the gases performs work. Brisance shows the ability of an explosive to demolish a solid object which is in direct contact or in the vicinity of the detonation wave impact. Thus, the shattering power of an explosive by brisance is different from the total work capacity of an explosive. Brisance shows the speed with which the explosive reaches its peak pressure. It has practical importance because it determines the effectiveness of an explosive in military applications such as fragmenting shells, bomb casings, grenades, and mines, as well as in imparting high velocities to the resulting fragments. Therefore, it may be related directly to the detonation pressure or detonation velocity, and indirectly to the heat of detonation.

6.3.1 Sand crushing test

Investigation of the reported sand crushing tests of pure energetic compounds and their mixtures with general formula $C_aH_bN_cO_d$, as well as aluminized explosives, has shown that the following correlation can be used to predict the brisance of an explosive with respect to TNT [175]:

$$\%f_{brisance,sand,TNT} = 85.5 + 4.812c + 2.556(d - a - b/2)$$
$$+ 19.69f^+_{brisance,sand} - 35.96f^-_{brisance,sand}, \qquad (6.17)$$

where $\%f_{brisance,sand,TNT}$ is the relative brisance with respect to TNT for the sand crushing test; $f^+_{brisance,sand}$ and $f^-_{brisance,sand}$ are correcting functions (positive and negative) for the values obtained on the basis of the elemental composition. For example, the value of $\%f_{brisance,sand,TNT}$ for tetryl ($C_7H_5N_5O_8$) can be calculated as follows:

$$\%f_{brisance,sand,TNT} = 85.5 + 4.812(5) + 2.556(8 - 7 - 5/2) + 19.69(1) - 35.96(0)$$
$$= 125.$$

The experimental value of $\%f_{brisance,sand,TNT}$ is in the range 113–123 [176].

6.3.1.1 Prediction of $f^+_{brisance,sand}$ and $f^-_{brisance,sand}$ for pure energetic materials

The values of $f^+_{brisance,sand}$ and $f^-_{brisance,sand}$ depend on the presence of certain molecular moieties in the molecular structure of pure $C_aH_bN_cO_d$ energetic compounds, and are specified in Table 6.2.

6.3.1.2 Sand crushing test for mixtures and aluminized explosives

For mixtures of different energetic components, if at least one of components follows the conditions given in Table 6.2, $f^+_{brisance,sand}$ and $f^-_{brisance,sand}$ are equal to the values specified by them. Thus, equation (6.17) can easily be calculated for explosive

Table 6.2: Values of $f^+_{brisance}$ and $f^-_{brisance}$.

Energetic compound or molecular fragment	$f^+_{brisance}$	$f^-_{brisance}$
$(CH_2ONO_2)_n$ or $C(CH_2ONO_2)_n$ or $(CH_2-NNO_2)_n$ or $(-HN-NO_2)_n$, where $n \le 4$ and aromatic $-N(NO_2)-$	1.0	—
O (or N) ‖ ⋀ `N`	—	1.0
Only $-ONO_2$ along with $-COO-$	—	2.0
More than one $-C-O-C-$	—	1.5
Nitramine group	—	1.0

mixtures from knowledge of the composition and the empirical formula. For example, the calculated $\%f_{brisance,sand,TNT}$ for 60/40 EDNA/TNT (Ednatol) with formula $C_{2.033}H_{3.281}N_{2.129}O_{2.657}$ containing 60 % ethylene dinitramine ($C_2H_6N_4O_4$) and 40 % TNT is

$$\%f_{brisance,sand,TNT} = 85.5 + 4.812(2.129) + 2.556(2.657 - 2.033 - 3.28/2)$$
$$+ 19.69(1) - 35.96(0)$$
$$= 112.$$

The measured value of $\%f_{brisance,sand,TNT}$ is in the range of 112 to 117 [176].

For aluminized explosives, the partial equilibrium prediction gives better results than the full equilibrium condition, because aluminum can interact with the detonation products. The following correlation can be used to predict the brisance of aluminized explosives with general formula $C_aH_bN_cO_dAl_e$ as

$$(\%f_{brisance,sand,TNT})_{aluminized\ explosive} = -42.87(d - a - b/2) + 146.71e, \qquad (6.18)$$

where $(\%f_{brisance,sand,TNT})_{aluminized\ explosive}$ is the relative brisance of the sand crushing test with respect to TNT for aluminized explosives. As is shown in equation (6.18), there is no need to use $f_{brisance,sand}^+$ and $f_{brisance,sand}^-$ for aluminized explosives. For example, torpex (45/37/18/RDX/TNT/Al) comprising 45 % RDX, %37 TNT and 18 % powdered aluminum with the empirical formula $C_{1.749}H_{2.031}N_{1.705}O_{2.194}Al_{0.667}$ is particularly useful in underwater munitions because the aluminum component has the effect of making the explosive pulse last longer, which increases the destructive power. The use of equation (6.18) gives $(\%f_{brisance,sand,TNT})_{aluminized\ explosive}$ as

$$(\%f_{brisance,sand,TNT})_{aluminized\ explosive} = -42.87(2.194 - 1.749 - 2.031/2) + 146.71(0.667)$$
$$= 122.$$

The measured value of $(\%f_{brisance,TNT})_{aluminized\ explosive}$ is also 122 [176].

6.3.2 Plate dent test

Since performing the plate dent test for every new explosive formulation can be tedious and time consuming, it is suitable to use a simple accurate relationship for the absolute dent depth estimation of pure and mixed organic explosives. The relative value of brisance of a desired explosive with respect to a dent depth of 6.706 mm produced by a standard TNT charge at 1.63 g/cm^3 is given as follows [177]:

$$\%f_{brisance,plate\ dent,TNT} = 14.911\delta, \qquad (6.19)$$

where $\%f_{\text{brisance,plate dent,TNT}}$ is the relative brisance with respect to TNT for the plate dent test; δ is the dent depth in mm. There is a good correlation between detonation pressure in kbar and δ [157]

$$P_{\text{det}} = 33.374\delta, \tag{6.20}$$

where P_{det} is in kbar. Equation (6.20) was derived when the absolute dent depth values δ were plotted against available experimental data p for 29 CHNO and CHNOClF-based explosives and one Ba-containing explosive [157]. It may give a reasonable estimate of the detonation pressure for most organic high explosives of military interest. Meanwhile, it does not hold true for metal-loaded explosives, such as lead- and tungsten-containing compositions. It was shown that there is a linear correlation between the Gurney velocity of a particular explosive obtained from the cylinder test and the corresponding dent depth divided by the density as follows [178]:

$$\sqrt{2E_G} = \frac{0.5935\delta^{0.8}}{\rho_0} + 0.697, \tag{6.21}$$

where $\sqrt{2E_G}$ is in km/s and ρ_0 is in g/cm^3. Equation (6.21) can be applied only for explosives containing C–H–N–O and may deviate for compositions containing metal additives. Thus, it is important to use δ for prediction of $\%f_{\text{brisance,plate dent,TNT}}$, $\sqrt{2E_G}$ and P_{det} through equations (6.19)–(6.21). Several approaches have been introduced for prediction of δ that are demonstrated in the following sections.

6.3.2.1 Heat of formation and elemental composition

It is shown that the loading density, the condensed phase heat of formation and the elemental composition are the most important parameters needed in order to accurately predict the dent depth produced on 1018 cold-rolled steel by a detonating organic explosive. Thus, the dent depth (in mm) can be obtained by these parameters as [178]:

$$\begin{aligned}
\delta = \exp\Big[&1.395 + 5.28 \times 10^{-4}\Delta_f H^\theta(\text{c}) + 0.367\rho_0^2 \\
&- \frac{21.871}{\text{formula weight of explosive}} \\
&- 8.93 \times 10^{-2}a - 3.75 \times 10^{-2}c + 6.71 \times 10^{-2}d\Big],
\end{aligned} \tag{6.22}$$

where $\Delta_f H^\theta(\text{c})$ is in kJ/mol and ρ_0 is in g/cm^3. For example, PBX-9404 has the composition 94/3/3 HMX/nitrocellulose/tris-β-chloroethyl phosphate ($C_{1.40}H_{2.75}N_{2.57}O_{2.69}Cl_{0.03}P_{0.01}$) with $\Delta_f H^\theta(\text{c}) = 0.33$ kJ/mol and $\rho_0 = 1.840$ g/cm^3. The use of equation (6.22) gives the dent depth data for PBX-9404 as follows

$$\delta = \exp\left[1.395 + 5.28 \times 10^{-4} \times 0.33 + 0.367(1.840)^2\right.$$

$$-\frac{21.871}{1.40 \times 12.01 + 2.75 \times 1.01 + 2.57 \times 14.01 + 2.69 \times 16.00 + 0.03 \times 35.45 + 0.01 \times 30.97}$$

$$\left. - 8.93 \times 10^{-2} \times 1.40 - 3.75 \times 10^{-2} \times 2.57 + 6.71 \times 10^{-2} \times 2.69\right]$$

$$= 10.782 \, \text{mm}.$$

The calculated value of δ is close to the measured data, i. e. 11.049 mm [157]. The calculated depth data for PBX-9404 can also be used to assess the relative brisance with respect to TNT, the detonation pressure and the Gurney velocity through equations (6.19) to (6.21) as follows:

$$\%f_{brisance,plate\,dent,TNT} = 14.911 \times 10.782 = 160,$$

$$P_{det} = 33.374 \times 10.782 = 360 \, \text{kbar},$$

$$\sqrt{2E_G} = \frac{0.5935 \times 10.782^{0.8}}{1.84} + 0.697 = 2.86 \, \text{km/s}.$$

It should be mentioned that the measured Gurnery velocity of PBX-9404 is 2.90 km/s [9].

6.3.2.2 The use of the "H_2O–CO_2 arbitrary" decomposition assumption

It is shown that the number of moles of gaseous detonation products per gram of explosive (n'_{gas}) and the average molecular weight of the gaseous products in g/mol decomposition products ($\bar{M}_{wgas}$) based on equation (1.10) are used to obtain the dent depth by [179]

$$\delta = -8.564 + 2301(n'_{gas}\rho_0)^2 + 0.3923\bar{M}_{wgas}, \tag{6.23}$$

where ρ_0 is in g/cm^3. For example, the values of n'_{gas} and $\bar{M}_{wgas}$ for HMX are calculated in Section 3.4.1.1, which are $n'_{gas} = 0.03378$ and $\bar{M}_{wgas} = 27.21$. The use of these data in equation (6.23) at $\rho_0 = 1.73$ g/cm^3 gives

$$\delta = -8.564 + 2301(0.03378 \times 1.73)^2 + 0.3923 \times 27.21 = 9.968 \, \text{mm}.$$

The calculated dent depth is close to experimental value, i. e. 10.084 mm [157].

6.3.2.3 The specific impulse and the oxygen balance

The specific impulse shows a measure of how effectively a rocket uses propellant or a jet engine uses fuel. It is defined as the total impulse (or change in momentum) delivered per unit of propellant consumed. The specific impulse may be calculated by the following correlation [180]:

$$I_{SP} = \left(-4.459 + 121.81n'_{gas} + 1.123Q_{det}[H_2O(g)]\right)^{0.5}, \tag{6.24}$$

where I_{SP} is the specific impulse in N.s/g, n'_{gas} is defined in the previous section and $Q_{det}[H_2O(g)]$ with unit kJ/g is the heat of detonation if H_2O is given as a gaseous detonation product. Two variables n'_{gas} and $Q_{det}[H_2O(g)]$ are calculated according to equations (1.5) and (1.10).

The oxygen balance indicates the degree to which an explosive can be oxidized. The parameter of the oxygen balance relative to CO_2 (Ω) for an explosive with general formula $C_aH_bN_cO_dCl_eF_f$ is defined as:

$$\Omega(\%) = \frac{32[0.5d + 0.25e + 0.25f - a - 0.25b]}{\text{formulaweight of explosive}} \times 100. \qquad (6.25)$$

It was indicated that the dent depth can be calculated by the oxygen balance and the specific impulse as follows [180]:

$$\delta = \exp[0.525 + 1.745\ln(\rho_0) + 0.350I_{SP} + 3.41 \times 10^{-3}\Omega(\%)], \qquad (6.26)$$

where ρ_0 is in g/cm^3 and I_{SP} is the specific impulse in N.s/g. For example, the calculated values of n'_{gas} and $Q_{det}[H_2O(g)]$ for HMX are 0.03378 (Section 3.4.1.1) and 6.18 kJ/g (Section 1.2.1). The use of these data in equation (6.24) gives:

$$I_{SP} = (-4.459 + 121.81 \times 0.03378 + 1.123 \times 6.18)^{0.5} = 2.57 \text{ N.s/g.}$$

Since chemical formula of HMX is also $C_4H_8N_8O_8$, equation (6.25) provides:

$$\Omega(\%) = \frac{32[0.5 \times 8 + 0.25 \times 0 + 0.25 \times 0 - 4 - 0.25 \times 8]}{4 \times 12.01 + 8 \times 1.01 + 8 \times 14.01 + 8 \times 16.00} \times 100 = -21.62.$$

The use of the calculated I_{SP} and $\Omega(\%)$ in equation (6.26) for HMX at $\rho_0 = 1.73\,g/cm^3$ gives:

$$\delta = \exp[0.525 + 1.745\ln(1.73) + 0.350 \times 2.57 + 3.41 \times 10^{-3} \times (-21.62)] = 10.040 \text{ mm.}$$

The calculated dent depth is close to the measured value, i. e. 10.084 mm [157].

Summary

This chapter has reviewed several empirical methods for the prediction of the power and brisance of energetic compounds. The methods which have been introduced for the assessment of the power were based on the Trauzl lead block and the ballistic mortar tests. Due to the existence of a large amount of experimental data from the Trauzl lead block test in the open literature, several approaches are available for the assessment of the power of an energetic compound using this test. One model has been introduced for the prediction of the brisance of energetic compounds and aluminized explosives measured by the sand crushing test. Three different methods have also been introduced to estimate brisance of high explosives by the plate dent test.

Questions and problems

> **i** The necessary information for some problems are given in the Appendix.

(1) Calculate the volume of explosion gases for butane-1,2,4-triyl trinitrate ($C_4H_7N_3O_9$).

(2) If the values of $Q_{det}[H_2O(l)]$ and I_{SP} for ammonium picrate are 3.046 kJ/g and 2.143 N s/g, respectively, calculate $\%f_{Trauzl,TNT}$ using equation (6.9).

(3) If {3-[3-(nitrooxy)-2,2-bis[(nitrooxy)methyl]propoxy]-2,2-bis[(nitrooxy)methyl] propyl} ($C_{10}H_{16}N_6O_{19}$) has $Q_{det}[H_2O(l)]$ = 6.629 kJ/g, calculate $\%f_{Trauzl,TNT}$ using equation (6.10).

(4) Calculate $\%f_{Trauzl,TNT}$ for 1,3,3-trinitroazetidine (TNAZ) and 2,4,6,8,10,12-hexa-nitro-2,4,6,8,10,12-hexaazaisowurtzitane (CL-20)
   (a) using equation (6.11) if the gas phase heat of formation of TNAZ and CL-20 are 127.8 and 676.0 kJ/mol, respectively;
   (b) using equation (6.12) if the condensed phase heat of formation of TNAZ and CL-20 are 14.48 and 414.22 kJ/mol, respectively.

(5) Using equation (6.13), calculate $\%f_{Trauzl,TNT}$ for AN/TNT (80/20).

(6) Use equation (6.15) to calculate $\%f_{ballistic\ mortar,TNT}$ for 1,1,1,3,5,5,5-Heptanitro-pentane ($C_5H_5N_7O_{14}$).

(7) Use equation (6.17) to calculate $\%f_{brisance,TNT}$ for 30/70 TNT/Tetryl (Tetrytol) ($C_{2.632}H_{1.88}N_{1.616}O_{2.744}$).

(8) Use equation (6.18) to calculate $\%f_{brisance,TNT}$ for 29/49/22 TNT/HMX/Al (HTA-3) ($C_{1.801}H_{2.016}N_{1.664}O_{2.192}Al_{0.667}$).

(9) Consider 93.9/2.3/3.8 RDX/DOP/PS with chemical formula $C_{1.698}H_{3.05}N_{2.538}O_{2.562}$, $\Delta_f H^\theta(c)$ = 22.30 kJ/mol and ρ_0 = 1.713 g/cm³. Use equation (6.22) to calculate the dent depth data for 93.9/2.3/3.8 RDX/DOP/PS.

(10) Use equation (6.23) to calculate the dent depth data for 70/30 RDX/TNT with chemical formula $C_{1.87}H_{2.56}N_{2.29}O_{2.68}$.

(11) Use equations (1.5), (1.10), (6.24) and (6.25) to calculate the dent depth data for 54.7/45.3 PETN/TNT with chemical formula $C_{2.258}H_{2.379}N_{1.289}O_{3.27}$ as well as $\Delta_f H^\theta(c)$ = −106.5 kJ/mol and ρ_0 = 1.655 g/cm³.

> **i** For answers and solutions, please see p. 120

7 Underwater detonation (explosion)

Underwater detonation (explosion) is important for some industrial applications such as blasting cut of old warships, blasting drill and decoupled charge of underwater blast. Distribution forms of energy of explosives reacting underwater are one of the main indexes evaluating the explosives output performance. Thus, there is a difference in energy between the explosion in air and in water. **Shock wave energy**, **bubble energy**, and **heat energy loss** due to the compression of the aqueous medium are three parts of the conversion of chemical energy of an explosive, which is detonated underwater. For the design of explosives with different underwater energy output structures, the proportion of the three parts of energy distribution and attenuation variation has to be taken into account when these energies propagate in water.

Al particles are widely used as additives in explosives for the underwater detonation because they can reduce the decay of the shock wave pressure, increase the bubble energy in underwater weapons, and influence the underwater warhead performance. Three common secondary explosives including TNT, RDX and HMX are usually employed as matrix explosives in underwater weapon warheads. AP is also often added to aluminized explosives for enhancement of the oxidizability of the detonation products. Since there is the physical separation of fuel and oxidizer, secondary reactions occur between detonation products and the chemical reaction zone spreads for composite explosives containing Al and/or AP. The partial equilibrium or consumption of a fraction of Al/AP should be considered because Al/AP composite explosives cannot be described by the steady-state detonation calculations [27, 39]. The modeling of nonideal behavior of Al/AP composite explosives is much more complicated because it requires incorporation of reaction kinetics into fluid-flow equations (e. g. Wood and Kirkwood model). It is important to know the exact relationship between the damage effect of the aluminized explosive and the underwater energy output structure for optimizing the design of charge of underwater weapons.

7.1 Measurement of shock wave energy and bubble energy

The underwater detonation can determine the strength of the explosives on the basis of measurable forms of energy released by the shock wave energy and the bubble energy. The bubble energy is the main part of the chemical energy of the explosive. It is formed upon the propagation of a shock wave through the water. For assessment of the effectiveness of explosives, the peak pressure of the shock wave, bubble period, attenuation time constant, shock wave impulse, energy, and energy flux density are studied in the underwater detonation [52]. The underwater detonation performance is important in the weapon design and the target damage effects for military explosives.

https://doi.org/10.1515/9783110677652-007

Different parameters have been studied on characteristics of the explosives in the underwater explosion that include initiation position, manner and aluminum-oxygen ratio compared with TNT [181].

Different types of testing environments are used for determination of the explosive strength from the underwater detonation test, e. g. natural lakes or seas as well as water-filled tanks of different sizes. The testing conditions have not been standardized yet. Since the charge mass of explosive may influence the test results, it is suitable to use a large volume of water for testing of explosive charges with greater mass. Therefore, the charge mass depends on the used tank size. The detonation of an explosive charge is usually done by means of a detonator, or via a booster, at a defined depth under the water surface. The shock wave time profile and the bubble pulse period, which are used to determine the shock wave energy and the bubble energy, respectively, are recorded at a specified distance from the explosive charge. The total energy of the explosive is determined by the sum of the shock wave energy and the bubble energy.

The underwater detonation test often contains several components:

(a) *Fixing the explosive charge* – A suitable tool is applied at a defined depth under the water surface.

(b) *Pressure gauge* – A gauge is fixed at a known distance from the explosive charge for measurement of the shock wave pressure. It is connected to a charge amplifier and oscilloscope. The piezoelectric pressure can be used for the shock wave pressure measurements.

(c) *Standard electric detonator* – A standard electric detonator, with or without the use of a booster, is usually used to initiate the explosive charge.

Figure 7.1 shows a schematic diagram for measuring the underwater explosion performance, which has been used for 1 kg charge of the RDX/AP-based aluminized explo-

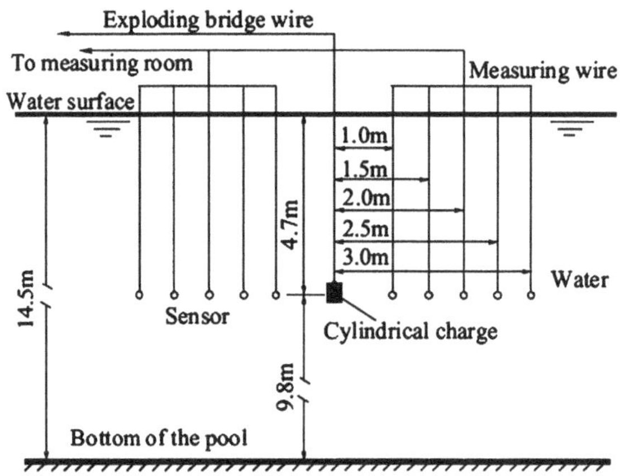

Figure 7.1: A schematic diagram for measuring the underwater explosion performance.

sives [181] and HMX-based aluminized explosives [182]. The diagram depicts a cylindrical water pool with a diameter of 85.0 m and a depth of 14.5 m. Cylindrical charges of the RDX/AP-based aluminized explosives are detonated in tap water at a depth of 4.7 m. Ten sensors are placed at distances 1.0 m, 1.5 m, 2.0 m, 2.5 m, and 3.0 m from the center of the explosive and facing the shock front. The length l and diameter d of the cylindrical charges ranged from 79.62 mm to 86.64 mm where the l/d ratios ranged from 1:1 to 1.2:1. Figure 7.2 gives a schematic diagram of the main charge as well as an initiation chain, which consists of a no. 8 electric detonator and a 1:1 right cylindrical pressed JH-14 booster of 10 g, Thus, the charges of the RDX/AP-based aluminized explosives are initiated by an initiation chain. It should be mentioned that the JH-14 booster has the composition 96.5 % RDX and 3.5 % fluororubber and graphite where its density and detonation velocity are 1.738 g/cm^3 and 8428 m/s, respectively.

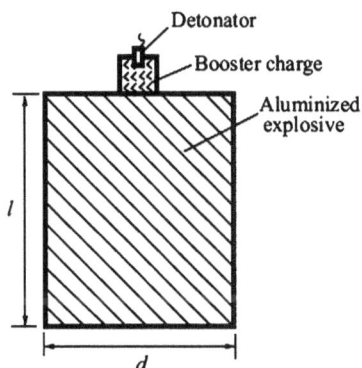

Figure 7.2: A schematic diagram of the main charge and an initiation chain.

The setup of Figure 7.1 can be used to measure the shock wave energy and the bubble energy that appear as the energy released by the detonation of the explosive charge underwater. The measured shock wave pressure-time dependency can determine the shock wave energy at a definite distance from the explosive charge as [183]:

$$E_S = \frac{4\pi r^2}{m v_W \rho_W} \int_0^{6.7\tau} p^2 dt, \tag{7.1}$$

where E_S is the shock wave energy; r is the distance between the pressure gauge and the charge; p is the shock wave pressure; m is the mass of explosive charge; v_W is the sound velocity in water or the acoustic velocity at the depth where the charge is positioned with allowance for the ambient water temperature; ρ_W is the density of water at the gauge location site; and τ is the representative time of the process, that is, the time during which the pressure signal recorded drops from its maximum (at the front) to $P_m/e \approx 0.37 P_m$ where P_m is the peak pressure of the shock wave.

The bubble pulse period can determine the bubble energy, which appears as the energy released by the detonation of the explosive charge underwater [184]:

$$E_b = (\sqrt{1 + 4CT_b} - 1)^3/8C^3k_1^3W,$$ (7.2)

$$k_1 = 1.135\rho_W^{1/2}/P_h^{5/6},$$ (7.3)

$$T_b = am^{1/3} + bm^{2/3},$$ (7.4)

$$C = \frac{b}{a^2},$$ (7.5)

where E_b denotes the gas bubble energy per kg explosives at the measuring point, MJ kg^{-1}; m is the charge weight of explosive (kg); ρ_W is the density of water (kg m^{-3}); T_b is the first pulsation period (s); P_h is the total hydrostatic pressure at the charge depth (including atmospheric pressure) (Pa). Two constants a and b describe the influence of the boundaries on the bubble period, such as used pools, and a certain hydrostatic pressure. A least-square fit to T_b and $m^{1/3}$ data can determine the values of a and b. Equation (7.4) is used to determine C, which is the adjustable parameter of boundary effects. Different masses (6 g to 10 g) of RDX are usually tested in the same condition in order to obtain C where the experimental data are fitted according to equation (7.3).

7.2 Assessment of the performance in the underwater explosion

For variations of the shock wave energy and the bubble energy with changes of the Al/O ratio in TNT/RDX/Al formulations [185], it was indicated that shock wave energy initially rises with increasing of the Al/O ratio up to a maximum at the Al/O ratio of 0.4, and then gradually decreases. In contrast, the bubble energy increases continuously in these formulations. For RDX/Al explosives, the same trend exists after performing a set of underwater detonation experiments [186]. The simulation of detonation of the Australian-made PBX-115 (43/25/20/12 AP/Al/RDX/HTPB) for its mid-scale underwater test has shown that the detonation velocity and critical diameter are sensitive to the assumed AP decomposition rate [187]. The effects of two particle sizes of aluminum (40 mm and 72 mm) on the detonation energy have also been studied for simulation of the gas bubble expansion process of PBXW-15 underwater [188]. The energy output of RDX/AP-based aluminized explosives of various formulations provides a theoretical guide for the design of RDX/AP-based aluminized explosives by recognizing the energy release rule, controlling the break process, increasing the energy efficiency, and improving blast effects [181]. The detonation and underwater detonation experiments on the performance of six HMX-based aluminized explosives have shown that when the Al/O mass ratio increases, the shock wave peak pressure increases and then decreases [182]. The effect of the charge construction on energy transmission has been investigated and the shock propagation rule underwater has been performed elsewhere [181, 182].

Assessment of the underwater performance of different formulations of aluminized explosives is essential for improving the design and widening the application of explosive formulations. Since predictions of the shock wave energy and the bubble energy require expensive experimental data [183], it is important to have suitable theoretical methods for their assessments.

The studies of the shock wave energy and the bubble energy of different composite explosives including compositions Al/AP/NM, RDX/Al/AP/Wax/Graphite, HMX/Al/wax/Graphite, RDX/wax/Al and RDX/Al with various $\frac{r}{m^{1/3}}$ have shown that the following correlations can be used for their estimations [189, 190]:

$$E_S = 0.1107c + 0.6942g - 1.261\frac{g}{d} - 0.0497\frac{r}{m^{\frac{1}{3}}} + 0.6648\rho_0, \tag{7.6}$$

$$E_b = 1.23 + 6.93f + 0.835a + 1.78g, \tag{7.7}$$

where E_S and E_b are the shock wave energy and the bubble energy in MJ/kg; r is the distance between the pressure gauge and the charge in meters; and m is the mass of the explosive charge in kilograms. For example, let us consider the use of Equations (7.6) and (7.7) for the calculation of the shock wave energy and the bubble energy of RDX/Al/AP/Wax/Graphite (20/30/45/3/2) with formula $C_{0.650}H_{2.513}N_{0.923}O_{2.072}Cl_{0.383}Al_{1.112}$ with $\frac{r}{m^{\frac{1}{3}}} = 3.0\,\text{m/kg}^{1/3}$ and $\rho_0 = 1.96\,\text{g/cm}^3$ as:

$$E_S = 0.1107c + 0.6942g - 1.261\frac{g}{d} - 0.0497\frac{r}{m^{\frac{1}{3}}} + 0.6648\rho_0$$

$$= 0.1107 \times 0.923 + 0.6942 \times 1.112 - 1.261 \times \frac{1.112}{2.072} - 0.0497 \times 3.0 + 0.6648 \times 1.96$$

$$= 1.35\,\text{MJ/kg},$$

$$E_b = 1.23 + 6.93f + 0.835a + 1.78g$$

$$= 1.23 + 6.93 \times 0.383 + 0.835 \times 0.650 + 1.78 \times 1.112 = 6.40\,\text{MJ/kg}.$$

The calculated shock wave energy and the bubble energy are close to the experimental values, which are 1.316 [181] and 6.328 [181], respectively.

It was also shown that Equations (7.6) and (7.7) can also be applied for pure and composite explosives, which do not contain Al/AP. For example, the composite explosive RDX/wax/Al (95/5/0) with formula $C_{1.639}H_{3.300}N_{2.566}O_{2.566}$ with $\frac{r}{m^{\frac{1}{3}}} = 5.5699\,\text{m/kg}^{1/3}$ and $\rho_0 = 1.60\,\text{g/cm}^3$ exhibits:

$$E_S = 0.1107c + 0.6942g - 1.261\frac{g}{d} - 0.0497\frac{r}{m^{\frac{1}{3}}} + 0.6648\rho_0$$

$$= 0.1107 \times 2.566 + 0.6942 \times 0 - 1.261 \times \frac{0}{2.566} - 0.0497 \times 5.5699 + 0.6648 \times 1.60$$

$$= 1.07\,\text{MJ/kg},$$

$$E_b = 1.23 + 6.93f + 0.835a + 1.78g$$
$$= 1.23 + 6.93 \times 0 + 0.835 \times 1.639 + 1.78 \times 0 = 2.60 \text{ MJ/kg}.$$

The calculated shock wave energy and the bubble energy are close to the experimental values, which are 1.044 [184] and 2.629 [184], respectively. Both experimental and calculated data confirm that the bubble energy of a composite with Al/AP is much higher than that of composite explosives without Al/AP.

Summary

Reliable predictive methods are important to assess the performance of composite explosives in underwater explosion before their manufacture and testing due to the difficulty, danger, and expenditure of experimental methods. They help the engineers to develop suitable composition having complementary properties of performance in underwater explosion. The shock wave energy and the bubble energy are the main parts of the chemical energy of the explosive that are formed by the underwater explosion upon the propagation of a shock wave through the water. The two simple methods are introduced for reliable prediction of the shock wave energy and the bubble energy of composite explosives containing Al and/or AP. The shock wave energy is based on the number of moles of nitrogen, oxygen, and Al atoms as well as loading density and the ratio $\frac{Rr}{m^{1/3}}$. Meanwhile, the bubble energy of composite Al/AP explosives depends on the number of moles of chlorine, carbon, and Al atoms. Equations (7.6) and (7.7) can also be applied for pure and composite explosives, which do not contain Al/AP.

Questions and problems

---

i   The necessary information for some problems are given in the Appendix.

---

(1) Calculate the shock wave energy of RDX/Al/wax (80/15/5) with chemical formula $C_{1.436}H_{2.895}N_{2.161}O_{2.161}Al_{0.556}$ with $\frac{r}{m^{\frac{1}{3}}} = 2.433 \text{ m/kg}^{1/3}$ and $\rho_0 = 1.76 \text{ g/cm}^3$.

(2) Calculate the bubble energy of HMX/Al/Wax/Graphite (85/10/3/2) with formula $C_{1.528}H_{2.736}N_{2.296}O_{2.296}Al_{0.371}$.

---

i   For answers and solutions, please see p. 121

Answers to questions and problems

Chapter 1

(1) According to equation (1.5), the more positive the value of the condensed phase heat of formation of an explosive is, the higher the value of the heat of detonation.

(2) $Q_{det}[H_2O(l)] = 4.94$ kJ/mol; $Q_{det}[H_2O(g)] = 4.48$ kJ/mol.

(3) **Ethyl nitrate:** $Q_{det}[H_2O(l)] = 3.63$ kJ/g from equation (1.16);
TATB: $Q_{det}[H_2O(l)] = 3.06$ kJ/g from equation (1.15).

(4) (a) $Q_{det}[H_2O(l)] = 2.763$ kJ/g.
   (b) $Q_{det}[H_2O(l)] = 4.835$ kJ/g from equation (1.10);
      $Q_{det}[H_2O(l)] = 2.901$ kJ/g from equation (1.11).
   (c) Equation (1.11) > equation (1.17) > equation (1.10).

(5) (a) $Q_{det}[H_2O(l)] = 2.988$ kJ/g (Dev = 0.16 kJ/g).
   (b) $Q_{det}[H_2O(l)] = 5.772$ kJ/g (Dev = 0.54 kJ/g).
   (c) $Q_{det}[H_2O(l)] = 4.364$ kJ/g (Dev = 0.39 kJ/g).
   (d) $Q_{det}[H_2O(l)] = 7.211$ kJ/g (Dev = 0.44 kJ/g).

(6) 6.88 kJ/g.

(7) 2.587 kJ/g.

(8) **NTO:** $Q_{det}[H_2O(l)] = 2.856$ kJ/g (Dev = 0.29 kJ/g) from equation (1.22);
      $Q_{det}[H_2O(l)] = 4.711$ kJ/g (Dev = −1.56 kJ/g) from Rice and Hare.
CL-20: $Q_{det}[H_2O(l)] = 6.525$ kJ/g (Dev = −0.21 kJ/g) from equation (1.22);
      $Q_{det}[H_2O(l)] = 6.945$ kJ/g (Dev = −0.63 kJ/g) from Rice and Hare.
FOX-7: $Q_{det}[H_2O(l)] = 5.019$ kJ/g (Dev = −0.58 kJ/g) from equation (1.22);
      $Q_{det}[H_2O(l)] = 5.971$ kJ/g (Dev = −1.53 kJ/g) from Rice and Hare.

(9) 3.79 kJ/g.

Chapter 2

(1) Since the heat of detonation is proportional to the detonation temperature, a more positive value for the condensed phase heat of formation of an explosive results in a higher value for the detonation temperature.

(2) 3554 K.

(3) 4689 K.

(4) 3965 K.

(5) 2637 K.

https://doi.org/10.1515/9783110677652-008

Chapter 3

(1) (a) Detonation velocities from equations (3.2) and (3.3) are 8.75 km/s and 8.80 km/s

    (b) Percent deviation of the calculated detonation velocities by equations (3.2) and (3.3) are −0.79 % and −0.23 %.

(2) 8.50 km/s.

(3) 8.81 km/s.

(4) $D_{det,max}$ from equations (3.6) and (3.7) are 7.56 and 7.36 km/s, respectively.

(5) 7.98 km/s.

(6) 7.46 km/s.

(7) 7.79 km/s.

Chapter 4

(1) (a) $n'_{gas} = 0.0410$ mol/g, $\bar{M}_{wgas} = 24.36$ g/mol, and $Q_{det}[H_2O(g)] = 4.26$ kJ/g.

    (b) 304 kbar, 6.1 %.

(2) 311 kbar.

(3) 301 kbar.

(4) 167 kbar.

(5) 187 kbar.

(6) 176 kbar.

(7) 246 kbar.

(8) 200 kbar and −5.1 %.

Chapter 5

(1) $R - R_0 = $ 6.0 mm $\rightarrow V_{cylinder\,wall} = 1.590$ km/s

    $R - R_0 = 19.0$ mm $\rightarrow V_{cylinder\,wall} = 1.769$ km/s.

(2) (a) $(\sqrt{2E_G})_{H-K} = 2.63$ km/s, $(\sqrt{2E_G})_{K-F} = 2.65$ km/s, and $\sqrt{2E_G} = 2.63$ km/s.

    (b) 2.63 km/s.

(3) 2.29 km/s.

(4) 2.36 km/s.

(5) 2.58 km/s.

Chapter 6

(1) 909 l/kg.

(2) 88.

(3) 150.

(4) (a) %$f_{Trauzl,TNT}$ for TNAZ and CL-20 are 185 and 176, respectively.

(b) %$f_{Trauzl,TNT}$ for TNAZ and CL-20 are 169 and 168, respectively.

(5) 148.

(6) 157.

(7) 112.

(8) 124.

(9) 8.933 mm.

(10) 9.012 mm.

(11) 8.209 mm.

Chapter 7

(1) 1.35 MJ/kg.

(2) 3.16 MJ/kg.

Appendix: Glossary of compound names and heats of formation for pure and composite explosives

Abbreviation	Full name or composition	Chemical formula	$\Delta_f H^\theta$(c) (kJ mol^{-1})
ABH	Azobis(2,2',4,4',6,6'-hexanitrobiphenyl)	$C_{24}H_6N_{14}O_{24}$	485.34 [39]
Alex 20	44/32/20/4 RDX/TNT/Al/Wax	$C_{1.783}H_{2.469}N_{1.613}O_{2.039}Al_{0.7335}$	−7.61
Alex 32	37/28/31/4 RDX/TNT/Al/Wax	$C_{1.647}H_{2.093}N_{1.365}O_{1.744}Al_{1.142}$	−9.33
Acrylonitrile/TNM (1/1.25 molar)	1/1.25 moles Acrylonitrile/Tetranitromethane	$C_{4.25}H_3N_6O_{10}$	208.31
AMATEX-20	42/20/38 AN/RDX/TNT	$C_{1.44}H_{1.38}N_{1.04}O_{1.54}(AN)_{0.53}$	−95.77
AMATEX-40	21/41/38 AN/RDX/TNT	$C_{1.73}H_{1.95}N_{1.61}O_{2.11}(AN)_{0.26}$	−197.49
AMATOL80/20	80/20 AN/TNT	$C_{0.62}H_{0.44}N_{0.26}O_{0.53}(AN)_1$	−371.25
AN	Ammonium Nitrate	NH_4NO_3 or $H_4N_2O_3$	−365.14 [9]
AN/ADNT (2/1 molar)	Ammonium Nitrate/1-Ammonium-3,5-dinitro-1,2,4-triazole (2/1 molar)	$C_{2.0}H_{12.0}N_{10.0}O_{10.0}$	−728.77
AN/ADNT/Al (2/1/2.66 molar)	Ammonium Nitrate/1-Ammonium-3,5-dinitro-1,2,4-triazole/Al (2/1/2.66 molar)	$C_{2.0}H_{12.0}N_{10.0}O_{10.0}Al_{2.66}$	−728.77
AN/ADNT/EDD (3/1/1 molar)	Ammonium Nitrate/1-Ammonium-3,5-dinitro-1,2,4-triazole/Ethylenediaminedinitrate (3/1/1 molar)	$C_{4.0}H_{26.0}N_{16.0}O_{19.0}$	−1747.87
AN/ADNT/NQ (1.38/1/1.83 molar)	Ammonium Nitrate/1-Ammonium-3,5-dinitro-1,2,4-triazole/1-Nitroguanidine (1.38/1/1.83 molar)	$C_{3.83}H_{16.84}N_{16.08}O_{11.80}$	−670.53
AN/ADNT/RDX (1.38/1/1.5 molar)	Ammonium Nitrate/1-Ammonium-3,5-dinitro-1,2,4-triazole/RDX (1.38/1/1.5 molar)	$C_{6.50}H_{18.52}N_{17.76}O_{17.14}$	−396.64
AN/ADNT/RDX (5/1/1 molar)	Ammonium Nitrate/1-Ammonium-3,5-dinitro-1,2,4-triazole/RDX (5/1/1 molar)	$C_{5.0}H_{30.0}N_{22.0}O_{25.0}$	−1755.36
AN/ADNT/RDX/Al (5/1/1/3.3 molar)	Ammonium Nitrate/1-Ammonium-3,5-dinitro-1,2,4-triazole/RDX/Al (5/1/1/3.3 molar)	$C_{5.0}H_{30.0}N_{22.0}O_{25.0}Al_{3.30}$	−1755.36
AN/ADNT/TATB (2/1/1.3 molar)	Ammonium Nitrate/1-Ammonium-3,5-dinitro-1,2,4-triazole/TATB (2/1/1.3 molar)	$C_{9.80}H_{19.80}N_{17.80}O_{17.80}$	−929.18

(continued)

Abbreviation	Full name or composition	Chemical formula	$\Delta_f H^{\theta}$ (c) (kJ mol^{-1})
ANFO 94.2/5.8	94.2/5.8 Ammonium Nitrate/Fuel Oil	$C_{0.418}H_{5.474}N_{2.353}O_{3.530}$	−428.97
ANTA	5-Amino-3-nitro-1H-1,2,4-triazole	$C_2H_3N_5O_2$	61.10 [191]
AP	Ammonium perchlorate	H_4NO_4Cl	−295.31 [80]
ARX-2002	61/20/19 RDX/Al/HTPB	$C_{2.21}H_{3.79}N_{1.65}O_{1.66}Al_{0.74}$	12.08
ARX-2010	41/20/19 RDX/Al/AP/HTPB	$C_{1.94}H_{3.25}N_{1.11}O_{1.12}Al_{0.74}AP_{0.17}$	−44.52
ATNI	Ammonium 2,4,5-Trinitroimidazole	$C_3H_4N_6O_6$	−86.02 [192]
bATetU	1,3-bis(5-amino-1H-tetrazol-1-yl)urea	$C_3H_6N_{12}O$	555.0 [193]
AN/Al (90/10)	–	$Al_{0.37}(AN)_{1.125}$ or $H_{4.5}N_{2.25}O_{3.37}Al_{0.37}$	−412.42
AN/Al (80/20)	–	$Al_{0.74}(AN)_1$ or $H_4N_2O_3Al_{0.74}$	−368.32
AN/Al (70/30)	–	$Al_{1.11}(AN)_{0.875}$ or $H_{3.5}N_{1.75}O_{2.62}Al_{1.11}$	−324.55
BTF	Benzotris(1,2,5-oxadiazole-1-oxide)	$C_6N_6O_6$	602.50 [39]
COMP A-3	91/9 RDX/WAX	$C_{1.87}H_{3.74}N_{2.46}O_{2.46}$	11.88 [39]
COMP B	63/36/1 RDX/TNT/wax	$C_{2.03}H_{2.64}N_{2.18}O_{2.67}$	5.36 [9]
Comp B, Grade A	63/36/1 RDX/TNT/wax	$C_{2.03}H_{2.64}N_{2.18}O_{2.65}$	7.74
COMP C-3	77/4/10/5/1/3 RDX/TNT/DNT/MNT/NC/Tetryl	$C_{1.90}H_{2.83}N_{2.34}O_{2.60}$	−26.99 [39]
COMP C-4	91/5.3/2.1/1.6 RDX/TNT/MNT/NC	$C_{1.82}H_{3.54}N_{2.46}O_{2.51}$	13.93 [39]
CPX-471	77/11.5/11.5 HMX/K10/polyNIMMO	$C_{1.87}H_{3.20}N_{2.28}O_{2.64}$	−10.84
CPX-472	77/11.5/11.5 HMX/K10/polyGLYN	$C_{1.77}H_{2.98}N_{2.30}O_{2.72}$	−12.13
CPX-473	77/11.5/11.5 HMX/BDNPAF/polyGLYN	$C_{1.60}H_{3.03}N_{2.32}O_{2.83}$	65.00
Cyclotol-50/50	50/50 RDX/TNT	$C_{2.22}H_{2.45}N_{2.01}O_{2.67}$	0.04
Cyclotol-60/40 (or COMP B-3)	60/40 RDX/TNT	$C_{2.04}H_{2.50}N_{2.15}O_{2.68}$	4.81 [9]
Cyclotol-65/35	65/35 RDX/TNT	$C_{1.96}H_{2.53}N_{2.22}O_{2.68}$	8.33
Cyclotol-70/30	70/30 RDX/TNT	$C_{1.87}H_{2.56}N_{2.29}O_{2.68}$	11.13
Cyclotol-75/25	75/25 RDX/TNT	$C_{1.78}H_{2.58}N_{2.36}O_{2.69}$	13.4 [9]
Cyclotol-77/23	77/23 RDX/TNT	$C_{1.75}H_{2.59}N_{2.38}O_{2.69}$	14.98
Cyclotol-78/22	78/22 RDX/TNT	$C_{1.73}H_{2.59}N_{2.40}O_{2.69}$	15.52

(continued)

Abbreviation	Full name or composition	Chemical formula	$\Delta_f H^\theta$ (c) (kJ mol^{-1})
DAAF	3,3′-Diamino-4,4′-azoxyfurazan	$C_4H_4N_8O_3$	443.0 [194]
DAAT	3,3′-Azobis(6-amino-1,2,4,5-tetrazine	$C_4H_4N_{12}$	862.0 [195]
DAAzF	3,3′-Diamino-4,4′-azofurazan	$C_4H_4N_8O_2$	536.0 [194]
DANP	1,3-Diazido-2-nitro-2-azapropane	$C_2H_4N_8O_2$	745.16 [196]
DATB	1,3-Diamino-2,4,6-trinitrobenzene	$C_6H_5N_5O_6$	-98.74 [39]
DAzBF	5,5′-Diazido-1H,1′H-3,3′-bi(1,2,4-triazole)	$C_4H_2N_{12}$	971.0 [75]
DCTri	4,5-Dicyano-2H-triazole	C_4HN_5	473.0 [193]
DEGN (or DEGDN)	Diethyleneglycol dinitrate	$C_4H_8N_2O_7$	-415.89
Destex	74.766/18.691/4.672/1.869 TNT/Al/Wax/Graphite	$C_{2.791}H_{2.3121}N_{0.987}O_{1.975}Al_{0.6930}$	-34.39
DIPAM (Dipicramide)	2,2′,4,4′,6,6′ -Hexanitro-[1,1-biphenyl]-3,3′-diamine	$C_{12}H_6N_8O_{12}$	-14.90 [197]
DIPAM (Dipicramide)	2,2′,4,4′,6,6′ -Hexanitro-[1,1-biphenyl]-3,3′-diamine	$C_{12}H_6N_8O_{12}$	-28.45 [39]
DAzBF	5,5′-Diazido-1H,1′H-3,3′-bi(1,2,4-triazole)	$C_4H_2N_{12}$	971.0 [75]
DCTri	4,5-Dicyano-2H-triazole	C_4HN_5	473.0 [193]
DDNP	6-Diazo-2,4-dinitro-2,4-cyclohexadien-1-one	$C_6H_2N_4O_5$	207.0 [3]
DEGN (or DEGDN)	Diethyleneglycol dinitrate	$C_4H_8N_2O_7$	-415.89
DNAF	4,4-Dinitro-3,3′-diazenofuroxan	$C_4N_8O_8$	665.70 [198]
DNDF	Dinitrodifuroxanyl	$C_4N_6O_8$	372.80 [199]
DNNC	1,3,5,5-Tetranitrohexahydropyrimidine	$C_4H_6N_6O_8$	0.40 [8]
1,2-DP	1,2-Bis(difluoroamino)propane	$C_3H_6N_2F_4$	-213.38 [200]
EARL-1	41.12/40.86/11.86/6.16 EDD/AN/RDX/A1	$C_{0.6}H_{4.57}N_{2.22}O_{3.18}Al_{0.23}$	-327.20
EDC-11	64/4/30/1/1 HMX/RDX/TNT/Wax/Trylene	$C_{1.986}H_{2.78}N_{2.23}O_{2.63}$	4.52
EDC-24	95/5 HMX/Wax	$C_{1.64}H_{3.29}N_{2.57}O_{2.57}$	18.28
EDD (or EDDN)	Ethylenediaminedinitrate	$C_2H_{10}N_4O_6$	-653.50 [1]
EDNA	N,N′-Dinitro-1,2-ethylene diamine (or Ethylene dinitramine)	$C_2H_6N_4O_4$	-103.82 [1]
EGDN (or Nitroglycol or NGC)	Ethyleneglycol dinitrate	$C_2H_4N_2O_6$	-242.78 [1]
EXP D	Ammonium picrate or Explosive D	$C_6H_6N_4O_7$	-393.30 [39]

(continued)

Abbreviation	Full name or composition	Chemical formula	$\Delta_f H^\theta$ (c) (kJ mol^{-1})
FEFO	Bis(2-fluoro-2,2-dinitroethyl)formal	$C_5H_6N_4O_{10}F_2$	−742.66 [80]
FM1	—	$C_{20}H_{30.2}N_{13.3}O_{33.2}F_{3.2}$	−668.18 [201]
FOX-7 (DADNE)	1,1-Diamino-2,2-dinitroethylene	$C_2H_4N_4O_4$	−130.0 [7]
FOX-12 (GUDN)	Guarnylureadinitramide	$C_2H_7N_7O_5$	−355.0 [202]
H6 (H-6)	44.15/29.31/20/6.54 RDX/TNT/Al/Paraffin	$C_{1.97}H_{2.77}N_{1.58}O_{1.97}Al_{0.74}$	−9.41
H$_2$BTA	Bistetrazolylamine	$C_2H_3N_9$	633.00 [203]
HBX-1	40/38/17/5 RDX/TNT/Al/Wax	$C_{2.07}H_{2.63}N_{1.58}O_{2.08}Al_{0.63}$	−9.48
HBX-3	31/29/35/5/0.5 RDX/TNT/AL/WAX/CaCl$_2$	$C_{1.66}H_{2.18}N_{1.21}O_{1.60}Al_{1.29}Ca_{0.005}Cl_{0.009}$	−8.71 [204]
HMX	Cyclotetramethylenetetranitramine	$C_4H_8N_8O_8$	74.98 [39]
HMX/Al (90/10)	—	$C_{1.216}H_{2.432}N_{2.432}O_{2.432}Al_{0.371}$	22.76
HMX/Al (80/20)	—	$C_{1.08}H_{2.16}N_{2.16}O_{2.16}Al_{0.715}$	20.21
HMX/Al (70/30)	—	$C_{0.944}H_{1.888}N_{1.888}O_{1.888}Al_{1.11}$	17.66
HMX/Al (60/40)	—	$C_{0.812}H_{1.624}N_{1.624}O_{1.624}Al_{1.483}$	15.19
HMX/Exon (90.54/9.46)	—	$C_{1.43}H_{2.61}N_{2.47}O_{2.47}F_{0.15}Cl_{0.10}$	−1026.80
HMX/AP/EDNP (51/20/29)	—	$C_{1.611}H_{3.6394}N_{1.811}O_{2.849}Cl_{0.170}$	−114.49
HMX/AP/PB (80.3/5.9/13.8)	—	$C_{2.10}H_{3.70}N_{2.17}O_{2.17}AP_{0.05}$	10.59
HMX/AP/PB (69/17/14)	—	$C_{1.97}H_{3.42}N_{1.86}O_{1.86}AP_{0.15}$	−20.09
HMX/AP/PB (57/29/14)	—	$C_{1.80}H_{3.09}N_{1.54}O_{1.54}AP_{0.25}$	−53.29
HMX/EDNP (71/29)	—	$C_{1.88}H_{3.498}N_{2.181}O_{2.708}$	−59.16
HMX/PB (86/14)	—	$C_{2.10}H_{3.70}N_{2.17}O_{2.17}$	25.42
HNAB	2,2',4,4',6,6'-Hexanitroazobenzene	$C_{12}H_4N_8O_{12}$	284.09 [39]
HNB	Hexanitrobenzene	$C_6N_6O_{12}$	65.69 [205]
HNE	1,1,1,2,2,2-Hexanitroethane	$C_2N_6O_{12}$	83.68 [199]
HNS	2,2',4,4',6,6'-Hexanitrostilbene	$C_{14}H_6N_6O_{12}$	78.26 [1]
1H-Tz	1H-Tetrazole	CH_2N_4	320.0 [206]
HyAz	Hydrazine Azide	H_5N_5	228.38 [207]

(continued)

Abbreviation	Full name or composition	Chemical formula	$\Delta_f H^\theta$ (c) (kJ mol^{-1})
HydTet	5,5'-hydrazotetrazole	$C_2H_4N_{10}$	414.0 [193]
LLM-105 (or ANPZO)	2,6-diamino-3,5-dinitropyrazine-1-oxide	$C_4H_4N_6O_5$	-41.42 [208]
LX-01	51.7/33.2/15.1 NM/TNM/1-Nitropropane	$C_{1.52}H_{3.73}N_{1.69}O_{3.39}$	-115.14
LX-04	85/15 HMX/Viton	$C_{5.485}H_{9.2229}N_8O_8F_{1.747}$	-89.96 [9]
LX-07	90/10 HMX/Viton	$C_{1.48}H_{2.62}N_{2.43}O_{2.43}F_{0.35}$	-51.46 [9]
LX-09	93/4.6/2.4 HMX/DNPA/FEFO	$C_{1.43}H_{2.74}N_{2.59}O_{2.72}F_{0.02}$	8.38 [9]
LX-10	95/5 HMX/Viton	$C_{1.42}H_{2.66}N_{2.57}O_{2.57}F_{0.17}$	-13.14 [9]
LX-11	80/20 HMX/Viton	$C_{1.61}H_{2.53}N_{2.16}O_{2.16}F_{0.70}$	-128.57 [9]
LX-14	95.5/4.5 HMX/Estane 5702-F1	$C_{1.52}H_{2.92}N_{2.59}O_{2.66}$	6.28 [39]
LX-15	95/5 HNS-I/Kel-F 800	$C_{3.05}H_{1.29}N_{1.27}O_{2.53}Cl_{0.04}F_{0.3}$	-18.16 [9]
LX-17	92.5/7.5 TATB/Kel-F 800	$C_{2.29}H_{2.18}N_{2.15}O_{2.15}Cl_{0.054}F_{0.2}$	-100.58 [9]
MEN-II	72.2/23.4/4.4 Nitromethane/Methanol/Ethylene diamine	$C_{2.06}H_{7.06}N_{1.33}O_{3.10}$	-310.87 [9]
LX-01	51.7/33.2/15.1 NM/TNM/1-Nitropropane	$C_{1.52}H_{3.73}N_{1.69}O_{3.39}$	-115.14
MINOL-2	40/40/20 AN/TNT/Al	$C_{1.23}H_{0.88}N_{0.53}O_{1.06}Al_{0.74}(AN)_{0.5}$	-194.26 [9]
NM	Nitromethane	$C_1H_3N_1O_2$	-112.97 [39]
NM/CT (50/50)	50/50 Nitromethane/Carbon Tetrachloride	$C_{1.14}H_{2.46}N_{0.82}O_{1.64}Cl_{1.30}$	-138.21
NM/Toluene (85.5/14.5)	–	$C_{2.503}H_{5.461}N_{1.4006}O_{2.8013}$	-156.85
NM/TNM (1/0.071 molar)	1/0.071 moles Nitromethane/Tetranitromethane	$C_{1.07}H_{3.0}N_{1.28}O_{2.57}$	-110.65
NM/TNM (1/0.25 molar)	1/0.25 moles Nitromethane/Tetranitromethane	$C_{1.25}H_{3.0}N_{2.0}O_{4.0}$	-103.76
NM/UP (60/40)	60/40 Nitromethane/UP; UP=90/10 CO(NH$_2$)$_2$HClO$_4$/H$_2$O	$C_{1.207}H_{4.5135}N_{1.432}O_{3.309}Cl_{0.2341}$	11.51
NN14Tri	3-Nitro-5-nitrimino-1,4H-1,2,4-triazole	$C_2H_2N_6O_4$	84.70 [209]
NN1Tri	4-Nitro-5-nitrimino-1H-1,2,4-triazole	$C_2H_2N_6O_4$	191.00 [209]
NONA	2,2',2'',4,4',4'',6,6',6''-Nonanitro-m-terphenyl	$C_{18}H_5N_9O_{18}$	114.64 [39]
NQ	Nitroguanidine	$CH_4N_4O_2$	-92.47 [9]
NQ/Estane (95/5)	–	$C_{1.17}H_{4.03}N_{3.66}O_{1.91}$	-103.86 [210]
(NQ)2Tz/Viton A (95/5)	95/5 3,6-bis-nitroguanyl tetrazine/Viton A	$C_{1.46}H_{2.09}N_{3.98}O_{1.33}F_{0.17}$	91.91 [210]

(continued)

Abbreviation	Full name or composition	Chemical formula	$\Delta_f H^\theta$ (c) (kJ mol^{-1})
NTO (or ONTA)	3-Nitro-1,2,4-triazole-5-one	$C_2H_2N_4O_3$	−100.78 [1]
5NTOxide	5-Nitrotetrazole-2N-oxide	CHN_5O_3	308.60 [178]
NTO/RDX/TNT (50/12/38)	NTO/RDX/TNT 50/12/38	$C_{2.10}H_{1.93}N_{2.36}O_{2.48}$	−45.33
NTO/TNT (60/40)	NTO/TNT 60/40	$C_{2.16}H_{1.80}N_{2.37}O_{2.44}$	−57.42
NTO/TNT (50/50)	NTO/TNT 50/50 (X-0489)	$C_{2.3097}H_{1.8694}N_{2.1980}O_{2.4740}$	−52.40
Octol-77.6/22.4	77.6/22.4 HMX/TNT	$C_{1.74}H_{2.59}N_{2.39}O_{2.69}$	13.54
Octol-76.3/23.7	76.3/23.7 HMX/TNT	$C_{1.76}H_{2.58}N_{2.37}O_{2.69}$	12.76
Octol-76/23	76.3/23.7 HMX/TNT	$C_{1.76}H_{2.58}N_{2.37}O_{2.69}$	12.76
Octol-75/25	75/25 HMX/TNT	$C_{1.78}H_{2.58}N_{2.36}O_{2.69}$	11.63 [9]
Octol-60/40	60/40 HMX/TNT	$C_{2.04}H_{2.50}N_{2.15}O_{2.68}$	4.14
PBX-9007	90/9.1/0.5/0.4 RDX/Polystyrene/DOP/Resin	$C_{1.97}H_{3.22}N_{2.43}O_{2.44}$	29.83 [39]
PBX-9010	90/10 RDX/Kel-F	$C_{1.39}H_{2.43}N_{2.43}O_{2.43}Cl_{0.09}F_{0.26}$	−32.93 [9]
PBX-9011	90/10 HMX/Estane	$C_{1.73}H_{3.18}N_{2.45}O_{2.61}$	−16.95 [39]
PBX-9205	92/6/2 RDX/Polystyrene/DOP	$C_{1.83}H_{3.14}N_{2.49}O_{2.51}$	24.31 [39]
PBX-9404	94/3/3 HMX/NC(12 %N)/CEF	$C_{1.40}H_{2.75}N_{2.57}O_{2.69}Cl_{0.03}$	3.37
PBX-9407	94/6 RDX/Exon 461	$C_{1.41}H_{2.66}N_{2.54}O_{2.54}Cl_{0.07}F_{0.09}$	3.39 [9]
PBX-9501	95/2.5/2.5 HMX/Estane/BDNPA-F	$C_{1.47}H_{2.86}N_{2.60}O_{2.69}$	9.62 [39]
PBX-9502	95/5 TATB/Kel-F 800	$C_{2.3}H_{2.23}N_{2.21}O_{2.21}Cl_{0.04}F_{0.13}$	−87.15 [9]
PBX-9503	15/80/5 HMX/TATB/KEL-F 800	$C_{2.16}H_{2.28}N_{2.26}O_{2.26}Cl_{0.038}$	−74.01 [9]
PBXC-9	75/20/5 HMX/Al/Viton	$C_{1.15}H_{2.14}N_{2.03}O_{2.03}F_{0.17}Al_{0.74}$	113.01
PBXC-116	86/14 RDX/Binder	$C_{1.968}H_{3.7463}N_{2.356}O_{2.4744}$	4.52
PBXC-117	71/17/12 RDX/Al/Binder	$C_{1.65}H_{3.1378}N_{1.946}O_{2.048}Al_{0.6303}$	−65.56
PBXC-119	82/18 HMX/Binder	$C_{1.817}H_{4.1073}N_{2.2149}O_{2.6880}$	18.28
PBXN-1	68/20/12 RDX/Al/Nylon	$C_{1.50}H_{2.86}N_{1.97}O_{1.97}Al_{0.74}$	−50.83
PBXN-109	67/20/13 RDX/Al/HTPB	$C_{1.86}H_{3.27}N_{1.81}O_{1.82}Al_{0.74}$	16.26
PBXW-115	20/43/25/12 RDX/AP/Al/HTPB	$C_{1.15}H_{3.36}N_{0.91}O_{2.01}Al_{0.93}$	−106.30

(continued)

Abbreviation	Full name or composition	Chemical formula	$\Delta_f H^\theta$ (c) (kJ mol^{-1})
Pentolite-50/50	50/50 TNT/PETN	$C_{2.33}H_{2.37}N_{1.29}O_{3.22}$	-100.01
PETN	Pentaerythritol tetranitrate	$C_5H_8N_4O_{12}$	-538.48 [9]
PF	1-Fluoro-2,4,6-trinitrobenzene	$C_6H_2N_3O_6F$	-224.72
Picramide	2,4,6-Trinitroaniline (TNA)	$C_6H_4N_4O_6$	-83.98 [1]
Picratol	52/48 Ammonium picrate/TNT	$C_{2.75}H_{2.32}N_{1.48}O_{2.75}$	-96.19
Picric acid (PA)	2,4,6-trinitrophenol	$C_6H_3N_3O_7$	-214.64
Pyridinediamine (PYX)	2,6-Bis(picrylamino)-3,5-dinitropyridine	$C_{17}H_7N_{11}O_{16}$	80.29 [8]
QM100R	–	$C_{7.3}H_{44.8}N_{18.9}O_{29}$	-271.12 [201]
RDX	Cyclomethylenetrinitramine	$C_3H_6N_6O_6$	61.55 [9]
RDX/Al (90/10)	–	$C_{1.215}H_{2.43}N_{2.43}O_{2.43}Al_{0.371}$	24.89
RDX/Al (80/20)	–	$C_{1.081}H_{2.161}N_{2.161}O_{2.161}Al_{0.715}$	22.13
RDX/Al (70/30)	–	$C_{0.945}H_{1.89}N_{1.89}O_{1.89}Al_{1.11}$	19.37
RDX/Al (60/40)	–	$C_{0.81}H_{1.62}N_{1.62}O_{1.62}Al_{1.483}$	16.61
RDX/Al (50/50)	–	$C_{0.675}H_{1.35}N_{1.35}O_{1.35}Al_{1.853}$	13.85
RDX/Exon (90.1/9.9)	–	$C_{1.44}H_{2.6}N_{2.44}O_{2.44}F_{0.17}Cl_{0.11}$	-195.48
RDX/TFNA (65/35)	–	$C_{1.54}H_{2.64}N_{2.2}O_{2.49}F_{0.44}$	-823.83
RF-02-32	82.2/9.33/8.47 HMX/K10/polyNIMMO	$C_{1.75}H_{3.08}N_{2.38}O_{2.66}$	-1.84
RX23AA		$H_{51.8}N_{29.1}O_{19.1}$	-115.90 [201]
RX26AF		$C_{20.4}H_{28.4}N_{25.9}O_{25.3}$	-86.19 [201]
RX36AH	51.32/43.68/5 HMX/BTF/Viton	$C_{1.870}H_{1.489}N_{2.426}O_{2.426}F_{0.171}$	80.84
RX41AB	95/5 K-6/Viton	$C_{1.34}H_{1.7}N_{2.41}O_{2.82}F_{0.17}$	-54.48
RX27AD	92.5/7.5 Tacot/Kel-f	$C_3H_{0.99}N_{1.91}O_{1.91}F_{0.2}Cl_{0.05}$	66.21
RX45AA	95/5 ANTA/Kel-f	$C_{1.57}H_{2-23}N_{3.68}O_{1.47}F_{0.13}Cl_{0.04}$	15.72
RX47AA	92.51/7.49 CL-14/Kel-f	$C_{2.31}H_{1.48}N_{2.17}O_{2.17}F_{0.20}Cl_{0.05}$	-12.08
RX48AA	92.37/7.63 ADNBF/Kel-f	$C_{2.45}H_{1.19}N_{1.92}O_{2.30}Cl_{0.06}$	14.33

(continued)

Abbreviation	Full name or composition	Chemical formula	$\Delta_f H^\theta$ (c) (kJ mol^{-1})
TACOT	tetranitrodibenzo-1,3 a,4,6 a-tetrazapentalene	$C_{12}H_4N_8O_8$	462.015 [77]
(TAG)$_2$(NQ)$_2$Tz	3,6-bis(nitroguanyl)-1,2,4,5-tetrazine bis (triaminoguanidinium)	$C_6H_{22}N_{24}O_4$	1255.0 [210]
TAT (TTA)	2,4,6-triazido-1,3,5-triazine (Cyanuric azide or cyanuric triazide)	C_3N_{12}	914.6 [199]
TATB	1,3,5-Triamino-2,4,6-trinitrobenzene	$C_6H_6N_6O_6$	−154.18 [9]
TATB/HMX/Kel-F (45/45/10)	—	$C_{1.88}H_{2.37}N_{2.26}O_{2.26}F_{0.28}Cl_{0.06}$	−478
Tetryl	N-Methyl-N-nitro-2,4,6-trinitroaniline	$C_7H_5N_5O_8$	19.54 [9]
TFENA	2,2,2-Trifluoroethylnitramine	$C_2H_3N_2O_2F_3$	−694.54
TFET	2,4,6-Trinitrophenyl-2,2,2-trifluoroethylnitramine	$C_8H_4N_5O_8F_3$	−576.8
TFNA	1,1,1-Trifluoro-3,5,5-trinitro-3-azahexane	$C_5H_7N_4O_6F_3$	−722.0 [77]
TNAZ	1,3,3-Trinitroazetidine	$C_3H_4N_4O_6$	36.4 [1]
TNETB/Al (90/10)	—	$C_{1.398}H_{1.398}N_{1.398}O_{3.263}Al_{0.371}$	−115.65
TNETB/Al (80/20)	—	$C_{1.243}H_{1.243}N_{1.243}O_{2.90}Al_{0.741}$	−102.80
TNETB/Al (70/30)	—	$C_{1.088}H_{1.088}N_{1.088}O_{2.538}Al_{1.112}$	−89.95
TNM	Tetranitromethane	CN_4O_8	38.50 [1]
TNT	2,4,6-Trinitrotoluene	$C_7H_5N_3O_6$	−63.2 [6]
TNT (Liquid)	—	$C_7H_5N_3O_6$	−37.37
TNTAB	Trinitrotriazidobenzene	$C_6N_{12}O_6$	1129.68 [39]
TNT/Al (95/5)	—	$C_{2.929}H_{2.092}N_{1.255}O_{2.511}Al_{0.185}$	−26.45
TNT/Al (90/10)	—	$C_{2.775}H_{1.982}N_{1.189}O_{2.378}Al_{0.370}$	−25.06
TNT/Al (89.4/10.6)	—	$C_{2.756}H_{1.969}N_{1.181}O_{2.362}Al_{0.393}$	−24.89
TNT/Al (85/15)	—	$C_{2.621}H_{1.872}N_{1.123}O_{2.246}Al_{0.555}$	−23.67
TNT/Al (80/20)	—	$C_{2.467}H_{1.762}N_{1.057}O_{2.114}Al_{0.741}$	−22.27
TNT/Al (78.3/21.7)	—	$C_{2.414}H_{1.724}N_{1.034}O_{2.069}Al_{0.804}$	−21.80
Toluene/Nitromethane(14.5/85.5)	—	$C_{2.503}H_{5.461}N_{1.4006}O_{2.8013}$	−160.71
Torpex	42/40/18 RDX/TNT/Al	$C_{1.8}H_{2.015}N_{1.663}O_{2.191}Al_{0.6674}$	−0.17
Tritonal	80/20 TNT/Al	$C_{2.465}H_{1.76}N_{1.06}O_{2.11}Al_{0.741}$	−23.64

Bibliography

[1] R. Meyer, J. Köhler, A. Homburg, Explosives, 6th ed., Wiley-VCH Verlag GmbH, Weinheim, Germany, 2007.

[2] T. M. Klapötke, Chemistry of High-Energy Materials, 5th ed., Walter de Gruyter GmbH Berlin/Boston, 2019.

[3] J. P. Agrawal, High Energy Materials: Propellants, Explosives and Pyrotechnics, Wiley-VCH Verlag GmbH & Co. KGaA, Wienheim, 2010.

[4] N. Kubota, Propellants and Explosives: Thermochemical Aspects of Combustion, 3rd ed., Wiley-VCH Verlag GmbH & Co. KGaA Weinheim, 2015.

[5] M. J. Kamlet, S. Jacobs, The chemistry of detonation. 1. A simple method for calculating detonation properties of CHNO explosives, Journal of Chemical Physics, 48 (1967) 23–35.

[6] P. E. Rouse Jr, Enthalpies of formation and calculated detonation properties of some thermally stable explosives, Journal of Chemical and Engineering Data, 21 (1976) 16–20.

[7] J. Akhavan, The Chemistry of Explosives, 3rd ed., Royal Society of Chemistry, Cambridge, UK, 2011.

[8] NIST Chemistry WebBook, https://webbook.nist.gov/chemistry/ provides chemical and physical property data for chemical species through the Internet, National Institute of Standards and Technology, Gaithersburg, Maryland.

[9] B. M. Dobratz, P. C. Crawford, LLNL Explosives Handbook: Properties of Chemical Explosives and Explosives Simulants, Lawrence Livermore National Lab., CA (USA), Livermore, CA, 1985.

[10] M. Sućeska, Calculation of detonation parameters by EXPLO5 computer program, Materials Science Forum, 465 (2004) 325–330.

[11] S. Bastea, L. E. Fried, K. R. Glaesemann, W. M. Howard, P. C. Souers, P. A. Vitello, CHEETAH 5.0 User's Manual Lawrence Livermore National Laboratory, Livermore, California, 2011.

[12] C. L. Mader, Numerical Modeling of Explosives and Propellants, 3rd ed., CRC Press, Florida, 2007.

[13] W. B. White, S. M. Johnson, G. B. Dantzig, Chemical equilibrium in complex mixtures, Journal of Chemical Physics, 28 (1958) 751–755.

[14] M. L. Hobbs, M. R. Baer, B. C. McGee, JCZS: An intermolecular potential database for performing accurate detonation and expansion calculations, Propellants, Explosives, Pyrotechnics, 24 (1999) 269–279.

[15] B. M. Rice, J. Hare, Predicting heats of detonation using quantum mechanical calculations, Thermochimica Acta, 384 (2002) 377–391.

[16] M. H. Keshavarz, Simple procedure for determining heats of detonation, Thermochimica Acta, 428 (2005) 95–99.

[17] A. Sikder, G. Maddala, J. Agrawal, H. Singh, Important aspects of behaviour of organic energetic compounds: a review, Journal of Hazardous Materials, 84 (2001) 1–26.

[18] P. F. Pagoria, G. S. Lee, A. R. Mitchell, R. D. Schmidt, A review of energetic materials synthesis, Thermochimica Acta, 384 (2002) 187–204.

[19] B. E. Poling, J. M. Prausnitz, O. C. John Paul, R. C. Reid, The Properties of Gases and Liquids, McGraw-Hill New York, 2001.

[20] O. V. Dorofeeva, M. A. Suntsova, Enthalpy of formation of CL-20, Computational and Theoretical Chemistry, 1057 (2015) 54–59.

[21] M. H. Keshavarz, Estimating heats of detonation and detonation velocities of aromatic energetic compounds, Propellants, Explosives, Pyrotechnics, 33 (2008) 448–453.

[22] M. H. Keshavarz, Predicting heats of detonation of explosives via specified detonation products and elemental composition, Indian Journal of Engineering & Materials Sciences, 14 (2007) 324–330.

https://doi.org/10.1515/9783110677652-010

[23] M. H. Keshavarz, Determining heats of detonation of non-aromatic energetic compounds without considering their heats of formation, Journal of Hazardous Materials, 142 (2007) 54–57.

[24] M. H. Keshavarz, Quick estimation of heats of detonation of aromatic energetic compounds from structural parameters, Journal of Hazardous Materials, 143 (2007) 549–554.

[25] M. H. Keshavarz, A simple way to predict heats of detonation of energetic compounds only from their molecular structures, Propellants, Explosives, Pyrotechnics, 37 (2012) 93–99.

[26] M. Rahmani, B. Ahmadi-Rudi, M. R. Mahmoodnejad, A. J. Senokesh, M. H. Keshavarz, Simple method for prediction of heat of explosion in double base and composite modified double base propellants, International Journal of Energetic Materials and Chemical Propulsion, 12 (2013) 41–60.

[27] M. H. Keshavarz, M. Kamalvand, M. Jafari, A. Zamani, An improved simple method for the calculation of the detonation performance of CHNOFCl, aluminized and ammonium nitrate explosives, Central European Journal of Energetic Materials, 13 (2016) 381–396.

[28] M. H. Keshavarz, A. Zamani, M. Shafiee, Predicting detonation performance of CHNOFCl and aluminized explosives, Propellants, Explosives, Pyrotechnics, 39 (2014) 749–754.

[29] M. H. Keshavarz, A. Zamani, A simple and reliable method for predicting the detonation velocity of CHNOFCl and aluminized explosives, Central European Journal of Energetic Materials, 12 (2015) 13–33.

[30] M. H. Keshavarz, R. Teimuri Mofrad, K. Esmail Poor, A. Shokrollahi, A. Zali, M. H. Yousefi, Determination of performance of non-ideal aluminized explosives, Journal of Hazardous Materials, 137 (2006) 83–87.

[31] M. H. Keshavarz, Simple correlation for predicting detonation velocity of ideal and non-ideal explosives, Journal of Hazardous Materials, 166 (2009) 762–769.

[32] M. H. Keshavarz, Predicting maximum attainable detonation velocity of CHNOF and aluminized explosives, Propellants, Explosives, Pyrotechnics, 37 (2012) 489–497.

[33] M. H. Keshavarz, Prediction of detonation performance of CHNO and CHNOAl explosives through molecular structure, Journal of Hazardous Materials, 166 (2009) 1296–1301.

[34] M. H. Keshavarz, A. Shokrolahi, H. R. Pouretedal, A new method to predict maximum attainable detonation pressure of ideal and aluminized energetic compounds, High Temperatures. High Pressures, 41 (2012) 349–365.

[35] A. H. Rezaei, M. H. Keshavarz, M. Kavosh Tehrani, S. M. R. Darbani, A. H. Farhadian, S. J. Mousavi, A. Mousaviazar, Approach for determination of detonation performance and aluminum percentage of aluminized-based explosives by laser-induced breakdown spectroscopy, Applied Optics, 55 (2016) 3233–3240.

[36] M. Jafari, M. H. Keshavarz, A simple method for calculating the detonation pressure of ideal and non-ideal explosives containing aluminum and ammonium nitrate, Central European Journal of Energetic Materials, 14 (2017) 966–983.

[37] M. H. Keshavarz, M. Jafari, R. Ebadpour, Simple method to calculate explosion temperature of ideal and non-ideal energetic compounds, Journal of Energetic Materials, https://doi.org/10.1080/07370652.2019.1679284.

[38] S. Grys, W. A. Trzciński, Calculation of combustion, explosion and detonation characteristics of energetic materials, Central European Journal of Energetic Materials, 7 (2010) 97–113.

[39] M. Hobbs, M. Baer, Calibrating the BKW-EOS with a large product species data base and measured CJ properties, in: Proc. of the 10th Symp. (International) on Detonation, ONR, 1993 p. 409.

[40] F. Gibson, M. Bowser, C. Summers, F. Scott, C. Mason, Use of an electro-optical method to determine detonation temperatures in high explosives, Journal of Applied Physics, 29 (1958) 628–632.

[41] V. Sil'vestrov, S. Bordzilovskii, S. Karakhanov, A. Plastinin, Temperature of the detonation front of an emulsion explosive, Combustion, Explosion, and Shock Waves, 51 (2015) 116–123.

[42] M. Tarasov, I. Karpenko, V. Sudovtsov, A. Tolshmyakov, Measuring the brightness temperature of a detonation front in a porous explosive, Combustion, Explosion, and Shock Waves, 43 (2007) 465–467.

[43] M. Sućeska, EXPLO5–Computer program for calculation of detonation parameters, in: Proc. of 32nd Int. Annual Conference of ICT, Karlsruhe, Germany, 2001.

[44] M. H. Keshavarz, Correlations for predicting detonation temperature of pure and mixed CNO and CHNO explosives, Indian Journal of Engineering and Materials Sciences, 12 (2005) 158–164.

[45] M. H. Keshavarz, M. Oftadeh, A new correlation for predicting the Chapman–Jouguet detonation pressure of CHNO explosives, High Temperatures. High Pressures, 34 (2002) 495–498.

[46] M. H. Keshavarz, M. Oftadeh, Two new correlations for predicting detonating power of CHNO explosives, Bulletin of the Korean Chemical Society, 24 (2003) 19–22.

[47] L. E. Fried, M. R. Manaa, P. F. Pagoria, R. L. Simpson, Design and synthesis of energetic materials 1, Annual Review of Materials Research, 31 (2001) 291–321.

[48] M. H. Keshavarz, H. R. Nazari, A simple method to assess detonation temperature without using any experimental data and computer code, Journal of Hazardous Materials, 133 (2006) 129–134.

[49] M. H. Keshavarz, Detonation temperature of high explosives from structural parameters, Journal of Hazardous Materials, 137 (2006) 1303–1308.

[50] C. Oommen, S. Jain, Ammonium nitrate: a promising rocket propellant oxidizer, Journal of Hazardous Materials, 67 (1999) 253–281.

[51] B. Wescott, D. S. Stewart, W. C. Davis, Equation of state and reaction rate for condensed-phase explosives, Journal of Applied Physics, 98 (2005) 053514.

[52] M. Sućeska, Test Methods for Explosives, Springer Science & Business Media, 1995.

[53] M. H. Keshavarz, A simple theoretical prediction of detonation velocities of non-ideal explosives only from elemental composition, in: P. B. Warey (Ed.) New Research on Hazardous Materials, Nova Science Publisher, Inc., New York, 2007, pp. 293–310.

[54] M. H. Keshavarz, Predicting detonation performance in non-ideal explosives by empirical methods, in: T. J. Janssen (Ed.) Explosive Materials: Classification, Composition and Properties, Nova Science Publishers, New York, 2011, pp. 179–201.

[55] S. Zeman, M. Jungová, Sensitivity and performance of energetic materials, Propellants, Explosives, Pyrotechnics, 41 (2016) 426–451.

[56] H. Shekhar, Studies on empirical approaches for estimation of detonation velocity of high explosives, Central European Journal of Energetic Materials, 9 (2012) 39–48.

[57] M. J. Kamlet, H. Hurwitz, Chemistry of detonations. IV. Evaluation of a simple predictional method for detonation velocities of CHNO explosives, Journal of Chemical Physics, 48 (1968) 3685–3692.

[58] M. Keshavarz, H. R. Pouretedal, Estimation of detonation velocity of CHNOFCl explosives, High Temperatures. High Pressures, 35 (2003) 593–600.

[59] M. H. Keshavarz, A simple approach for determining detonation velocity of high explosive at any loading density, Journal of Hazardous Materials, 121 (2005) 31–36.

[60] U. Nair, S. Asthana, A. S. Rao, B. Gandhe, Advances in high energy materials (review paper), Defence Science Journal, 60 (2010) 137–151.

[61] M. H. Keshavarz, R. T. Mofrad, R. F. Alamdari, M. H. Moghadas, A. R. Mostofizadeh, H. Sadeghi, Velocity of detonation at any initial density without using heat of formation of explosives, Journal of Hazardous Materials, 137 (2006) 1328–1332.

[62] M. Finger, E. Lee, F. Helm, B. Hayes, H. Hornig, R. McGuire, M. Kahara, M. Guidry, The effect of elemental composition on the detonation behavior of explosives, in: Sixth Symposium (International) on Detonation, 1976, pp. 710.

[63] L. Rothstein, R. Petersen, Predicting high explosive detonation velocities from their composition and structure, Propellants, Explosives, Pyrotechnics, 4 (1979) 56–60.

[64] L. R. Rothstein, Predicting high explosive detonation velocities from their composition and structure (II), Propellants, Explosives, Pyrotechnics, 6 (1981) 91–93.

[65] M. H. Keshavarz, Detonation velocity of pure and mixed CHNO explosives at maximum nominal density, Journal of Hazardous Materials, 141 (2007) 536–539.

[66] A. Elbeih, J. Pachmáň, S. Zeman, W. A. Trzciński, Z. Akštein, M. Sućeska, Thermal stability and detonation characteristics of pressed and elastic explosives on the basis of selected cyclic nitramines, Central European Journal of Energetic Materials, 7 (2010) 217–232.

[67] T. M. Klapötke, C. M. Sabaté, M. Rasp, Synthesis and properties of 5-nitrotetrazole derivatives as new energetic materials, Journal of Materials Chemistry, 19 (2009) 2240–2252.

[68] J. Boileau, C. Fauquignon, B. Hueber, Explosives, in: Ullmann's Encyclopedia of Industrial Chemistry, Wiley-VCH Verlag GmbH & Co. KGaA, Weinheim, Germany, 2011.

[69] M. B. Talawar, A. P. Agrawal, M. Anniyappan, D. S. Wani, M. K. Bansode, G. M. Gore, Primary explosives: Electrostatic discharge initiation, additive effect and its relation to thermal and explosive characteristics, Journal of Hazardous Materials, 137 (2006) 1074–1078.

[70] M. J. Kamlet, H. Hurwitz, Chemistry of detonations. IV. Evaluation of a simple predictional method for detonation velocities of C–H–N–O explosives, Journal of Chemical Physics, 48 (1968) 3685–3692.

[71] M. Jafari, M. H. Keshavarz, A. Zamani, S. Zakinejad, I. Alekaram, A novel method for assessment of the velocity of detonation for primary explosives, Propellants, Explosives, Pyrotechnics, 43 (2018) 342–347.

[72] R. Matyas, J. Pachman, Primary Explosives, Springer-Verlag Berlin, Heidelberg, New York, 2013.

[73] V. Sinditskii, V. Kolesov, V. Y. Egorshev, D. Patrikeev, O. Dorofeeva, Thermochemistry of cyclic acetone peroxides, Thermochimica Acta, 585 (2014) 10–15.

[74] B. Hariharanath, K. Chandrabhanu, A. Rajendran, M. Ravindran, C. Kartha, Detonator using nickel hydrazine nitrate as primary explosive, Defence Science Journal, 56 (2006) 383–389.

[75] A. A. Dippold, T. M. Klapötke, Nitrogen-Rich Bis-1,2,4-triazoles – A Comparative Study of Structural and Energetic Properties, Chemistry, An European Journal, 18 (2012) 16742–16753.

[76] A. E. Contini, A. J. Bellamy, L. N. Ahad, Taming the beast: measurement of the enthalpies of combustion and formation of triacetone triperoxide (TATP) and diacetone diperoxide (DADP) by oxygen bomb calorimetry, Propellants, Explosives, Pyrotechnics, 37 (2012) 320–328.

[77] M. Sućeska, EXPLO5 User's Guide, Version 6.02, Zagreb, Croatia, 2014, pp. 125.

[78] M. A. Ilyushin, I. V. Tselinsky, I. V. Shugalei, Environmentally friendly energetic materials for initiation devices, Central European Journal of Energetic Materials, 9 (2012) 293–327.

[79] T. O. Owolabi, Determination of the velocity of detonation of primary explosives using genetically optimized support vector regression, Propellants, Explosives, Pyrotechnics, (2019).

[80] P. W. Cooper, Explosives Engineering, Vch Pub, 1996.

[81] C. H. Johansson, P.-A. Persson, Detonics of High Explosives, Academic Press, illustrated edition 1970.

[82] D. Hardesty, J. Kennedy, Thermochemical estimation of explosive energy output, Combustion and Flame, 28 (1977) 45–59.

[83] M. J. Kamlet, J. M. Short, The chemistry of detonations. VI. A "Rule for Gamma" as a criterion for choice among conflicting detonation pressure measurements, Combustion and Flame, 38 (1980) 221–230.

[84] M. H. Keshavarz, H. R. Pouretedal, Predicting adiabatic exponent as one of the important factors in evaluating detonation performance, Indian Journal of Engineering and Materials Sciences, 13 (2006) 259.

[85] W. Davis, D. Venable, Pressure measurements for composition B-3, in: Fifth Symposium (International) on Detonation, 1970.

[86] M. J. Kamlet, J. E. Ablard, Chemistry of detonations. II. Buffered equilibria, Journal of Chemical Physics, 48 (1968) 36–42.

[87] M. J. Kamlet, C. Dickinson, Chemistry of detonations. III. Evaluation of the simplified calculational method for Chapman–Jouguet detonation pressures on the basis of available experimental information, Journal of Chemical Physics, 48 (1968) 43–50.

[88] L. Kazandjian, J. F. Danel, A Discussion of the Kamlet–Jacobs formula for the detonation pressure, Propellants, Explosives, Pyrotechnics, 31 (2006) 20–24.

[89] M. H. Keshavarz, H. R. Pouretedal, An empirical method for predicting detonation pressure of CHNOFCl explosives, Thermochimica Acta, 414 (2004) 203–208.

[90] M. H. Keshavarz, Simple determination of performance of explosives without using any experimental data, Journal of Hazardous Materials, 119 (2005) 25–29.

[91] M. Oftadeh, M. H. Keshavarz, R. Khodadadi, Prediction of the condensed phase enthalpy of formation of nitroaromatic compounds using the estimated gas phase enthalpies of formation by the PM3 and B3LYP methods, Central European Journal of Energetic Materials, 11 (2014) 143–156.

[92] M. H. Keshavarz, Reliable estimation of performance of explosives without considering their heat contents, Journal of Hazardous Materials, 147 (2007) 826–831.

[93] M. H. Keshavarz, Theoretical prediction of detonation pressure of CHNO high energy materials, Indian Journal of Engineering and Materials Sciences, 14 (2007) 77–80.

[94] Q. Zhang, Y. Chang, Prediction of detonation pressure and velocity of explosives with micrometer aluminum powders, Central European Journal of Energetic Materials, 9 (2012) 77–86.

[95] J. L. Gottfried, Laser-induced air shock from energetic materials (LASEM) method for estimating detonation performance: Challenges, successes and limitations, in: AIP Conference Proceedings, AIP Publishing, 2018, pp. 100014.

[96] J. L. Gottfried, T. M. Klapötke, T. G. Witkowski, Estimated Detonation Velocities for TKX-50, MAD-X1, BDNAPM, BTNPM, TKX-55, and DAAF using the Laser–induced Air Shock from Energetic Materials Technique, Propellants, Explosives, Pyrotechnics, 42 (2017) 353–359.

[97] J. L. Gottfried, D. K. Smith, C.-C. Wu, M. L. Pantoya, Improving the explosive performance of aluminum nanoparticles with aluminum iodate hexahydrate (AIH), Scientific Reports, 8 (2018) 8036.

[98] A. H. Rezaei, M. H. Keshavarz, M. Kavosh Tehrani, S. M. R. Darbani, Quantitative analysis for the determination of aluminum percentage and detonation performance of aluminized plastic bonded explosives by laser-induced breakdown spectroscopy, Laser Physics, 28 (2018) 065605.

[99] A. H. Rezaei, M. H. Keshavarz, M. Kavosh Tehrani, S. M. R. Darbani, Assessment of detonation performance and characteristics of 2,4,6-trinitrotoluene based melt cast explosives containing aluminum by laser induced breakdown spectroscopy, Central European Journal of Energetic Materials, 16 (2019) 3–20.

[100] J. L. Gottfried, Influence of exothermic chemical reactions on laser-induced shock waves, Physical Chemistry Chemical Physics, 16 (2014) 21452–21466.

[101] J. L. Gottfried, Laboratory-scale method for estimating explosive performance from laser-induced shock waves, Propellants, Explosives, Pyrotechnics, 40 (2015) 674–681.

[102] S. Roy, N. Jiang, H. U. Stauffer, J. B. Schmidt, W. D. Kulatilaka, T. R. Meyer, C. E. Bunker, J. R. Gord, Spatially and temporally resolved temperature and shock-speed measurements behind a laser-induced blast wave of energetic nanoparticles, Journal of Applied Physics, 113 (2013) 184310.

[103] W. Guo, X. Zheng, G. Yu, J. Zhao, Y. Zeng, C. Liu, Investigation of laser induced breakdown in liquid nitromethane using nanosecond shadowgraphy, Journal of Applied Physics, 120 (2016) 123301.

[104] S. A. Kalam, N. L. Murthy, P. Mathi, N. Kommu, A. K. Singh, S. V. Rao, Correlation of molecular, atomic emissions with detonation parameters in femtosecond and nanosecond LIBS plasma of high energy materials, Journal of Analytical Atomic Spectrometry, 32 (2017) 1535–1546.

[105] Y. A. Rezunkov, Laser reactive thrust. Review of research, Journal of Optical Technology, 74 (2007) 526–535.

[106] E. S. Collins, J. L. Gottfried, Laser-induced deflagration for the characterization of energetic materials, Propellants, Explosives, Pyrotechnics, 42 (2017) 592–602.

[107] J. L. Gottfried, E. J. Bukowski, Laser-shocked energetic materials with metal additives: evaluation of chemistry and detonation performance, Applied Optics, 56 (2017) B47–B57.

[108] D. W. Hahn, N. Omenetto, Laser-induced breakdown spectroscopy (LIBS), part I: review of basic diagnostics and plasma–particle interactions: still-challenging issues within the analytical plasma community, Applied Spectroscopy, 64 (2010) 335A–366A.

[109] D. W. Hahn, N. Omenetto, Laser-induced breakdown spectroscopy (LIBS), part II: review of instrumental and methodological approaches to material analysis and applications to different fields, Applied Spectroscopy, 66 (2012) 347–419.

[110] L. A. Skvortsov, Laser methods for detecting explosive residues on surfaces of distant objects, Quantum Electronics, 42 (2012) 1.

[111] M. R. Leahy-Hoppa, J. Miragliotta, R. Osiander, J. Burnett, Y. Dikmelik, C. McEnnis, J. B. Spicer, Ultrafast laser-based spectroscopy and sensing: applications in LIBS, CARS, and THz spectroscopy, Sensors, 10 (2010) 4342–4372.

[112] D. Rusak, B. Castle, B. Smith, J. Winefordner, Fundamentals and applications of laser-induced breakdown spectroscopy, Critical Reviews in Analytical Chemistry, 27 (1997) 257–290.

[113] P. Lucena, A. Doña, L. Tobaria, J. Laserna, New challenges and insights in the detection and spectral identification of organic explosives by laser induced breakdown spectroscopy, Spectrochimica Acta, Part B: Atomic Spectroscopy, 66 (2011) 12–20.

[114] C. G. Parigger, Atomic and molecular emissions in laser-induced breakdown spectroscopy, Spectrochimica Acta, Part B: Atomic Spectroscopy, 79 (2013) 4–16.

[115] Á. Fernández-Bravo, T. Delgado, P. Lucena, J. J. Laserna, Vibrational emission analysis of the CN molecules in laser-induced breakdown spectroscopy of organic compounds, Spectrochimica Acta, Part B: Atomic Spectroscopy, 89 (2013) 77–83.

[116] J. L. Gottfried, Laser-induced plasma chemistry of the explosive RDX with various metallic nanoparticles, Applied Optics, 51 (2012) B13–B21.

[117] V. Lazic, A. Palucci, S. Jovicevic, C. Poggi, E. Buono, Analysis of explosive and other organic residues by laser induced breakdown spectroscopy, Spectrochimica Acta, Part B: Atomic Spectroscopy, 64 (2009) 1028–1039.

[118] R. Lasheras, C. Bello-Galvez, E. Rodriguez-Celis, J. Anzano, Discrimination of organic solid materials by LIBS using methods of correlation and normalized coordinates, Journal of Hazardous Materials, 192 (2011) 704–713.

[119] M. Z. Martin, N. Labbé, T. G. Rials, S. D. Wullschleger, Analysis of preservative-treated wood by multivariate analysis of laser-induced breakdown spectroscopy spectra, Spectrochimica Acta, Part B: Atomic Spectroscopy, 60 (2005) 1179–1185.

[120] H. Fink, U. Panne, R. Niessner, Process analysis of recycled thermoplasts from consumer electronics by laser-induced plasma spectroscopy, Analytical Chemistry, 74 (2002) 4334–4342.

[121] D. Alamelu, A. Sarkar, S. Aggarwal, Laser-induced breakdown spectroscopy for simultaneous determination of Sm, Eu and Gd in aqueous solution, Talanta, 77 (2008) 256–261.

[122] N. K. Rai, A. K. Rai, A. Kumar, S. N. Thakur, Detection sensitivity of laser-induced breakdown spectroscopy for Cr II in liquid samples, Applied Optics, 47 (2008) G105–G111.

[123] S. Pandhija, N. Rai, A. K. Rai, S. N. Thakur, Contaminant concentration in environmental samples using LIBS and CF-LIBS, Applied Physics B, 98 (2010) 231–241.

[124] M. H. Keshavarz, H. Motamedoshariati, R. Moghayadnia, H. R. Nazari, J. Azarniamehraban, A new computer code to evaluate detonation performance of high explosives and their thermochemical properties, part I, Journal of Hazardous Materials, 172 (2009) 1218–1228.

[125] Q. Zhang, Y. Chang, A predictive method for the heat of explosion of non-ideal aluminized explosives, Central European Journal of Energetic Materials, 10 (2013) 541–554.

[126] Z. Q. Zhou, J. X. Nie, L. Zeng, Z. X. Jin, Q. J. Jiao, Effects of aluminum content on TNT detonation and aluminum combustion using electrical conductivity measurements, Propellants, Explosives, Pyrotechnics, 41 (2016) 84–91.

[127] V. Prakash, V. Phadke, R. Sinha, H. Singh, Influence of aluminium on performance of HTPB-based aluminised PBXs, Defence Science Journal, 54 (2004) 475.

[128] S. Mousavi, M. H. Farsani, S. Darbani, A. Mousaviazar, M. Soltanolkotabi, A. E. Majd, CN and C2 vibrational spectra analysis in molecular LIBS of organic materials, Applied Physics B, 122 (2016) 106.

[129] B. C. McGee, M. L. Hobbs, M. R. Baer, Exponential 6 Parameterization for the JCZ3-EOS, Sandia National Laboratories, Albuquerque, New Mexico, 1998.

[130] M. Held, Fragmentation warheads, in: J. Carleone (Ed.) Progress in Astronautics and Aeronautics, 1993, p. 387.

[131] D. Villano, F. Galliccia, Innovative technologies for controlled fragmentation warheads, Journal of Applied Mechanics, 80 (2013) 031704.

[132] R. W. Gurney, The Initial Velocities of Fragments from Bombs, Shell and Grenades, DTIC Document, 1943.

[133] P. C. Souers, J. W. Forbes, L. E. Fried, W. M. Howard, S. Anderson, S. Dawson, P. Vitello, R. Garza, Detonation energies from the cylinder test and CHEETAH V3.0, Propellants, Explosives, Pyrotechnics, 26 (2001) 180–190.

[134] J. M. Short, F. H. Helm, M. Finger, M. J. Kamlet, The chemistry of detonations. VII. A simplified method for predicting explosive performance in the cylinder test, Combustion and Flame, 43 (1981) 99–109.

[135] F. Fotouhi-Far, H. Bashiri, M. Hamadanian, M. H. Keshavarz, A new method for assessment of performing mechanical works of energetic compounds by the cylinder test, Zeitschrift für anorganische und allgemeine Chemie, 642 (2016) 1086–1090.

[136] C. M. Tarver, Condensed matter detonation: Theory and practice, in: Shock Waves Science and Technology Library, Vol. 6, Springer, 2012, pp. 339–372.

[137] G. I. Kanel, S. V. Razorenov, V. E. Fortov, Shock-Wave Phenomena and the Properties of Condensed Matter, Springer Science & Business Media, 2013.

[138] M. Cowperthwaite, W. Zwisler, Improvement and Modification to TIGER Code, SRI Final Report, Project PYU-1397 (2nd ed.)(January 1973), 1973.

[139] M. J. Kamlet, M. Finger, An alternative method for calculating Gurney velocities, Combustion and Flame, 34 (1979) 213–214.

[140] M. H. Keshavarz, A. Semnani, The simplest method for calculating energy output and Gurney velocity of explosives, Journal of Hazardous Materials, 131 (2006) 1–5.

[141] M. H. Keshavarz, New method for prediction of the Gurney energy of high explosives, Propellants, Explosives, Pyrotechnics, 33 (2008) 316–320.

[142] L. Türker, Thermobaric and enhanced blast explosives (TBX and EBX), Defence Technology, 12 (2016) 423–445.

[143] D. Frem, Estimating the metal acceleration ability of high explosives, Defence Technology, (2019).

[144] A. Rai, K. Park, L. Zhou, M. Zachariah, Understanding the mechanism of aluminium nanoparticle oxidation, Combustion Theory and Modelling, 10 (2006) 843–859.

[145] P. Vadhe, R. Pawar, R. Sinha, S. Asthana, A. S. Rao, Cast aluminized explosives, Combustion, Explosion, and Shock Waves, 44 (2008) 461–477.

[146] P. E. Anderson, P. Cook, A. Davis, K. Mychajlonka, The effect of binder systems on early aluminum reaction in detonations, Propellants, Explosives, Pyrotechnics, 38 (2013) 486–494.

[147] P. Anderson, P. Cook, W. Balas-Hummers, A. Davis, K. Mychajlonka, The detonation properties of combined effects explosives, in: Symposium Y – Advances in Energetic Materials Research, 2012, pp. mrsf11-1405-y1406-1403.

[148] V. W. Manner, S. J. Pemberton, J. A. Gunderson, T. J. Herrera, J. M. Lloyd, P. J. Salazar, P. Rae, B. C. Tappan, The role of aluminum in the detonation and post-detonation expansion of selected cast HMX-based explosives, Propellants, Explosives, Pyrotechnics, 37 (2012) 198–206.

[149] E. L. Baker, L. I. Stiel, W. Balas, C. Capellos, J. Pincay, Combined effects aluminized explosives modeling and development, International Journal of Energetic Materials and Chemical Propulsion, 14 (2015) 283–293.

[150] E. L. Baker, W. Balas, C. Capellos, J. Pincay, L. Stiel, Combined Effects Aluminized Explosives, Army Armament Research Developement and Engineering Center Picatinny Arsenal, 2010.

[151] E. Baker, D. Murphy, D. Suarez, C. Capellos, P. Cook, P. Anderson, E. Wrobel, L. Stiel, Recent Combined Effects Explosives Technology, Army Armament Research Developement and Engineering Center Picatinny Arsenal, 2010.

[152] D. Frem, A simple relationship for the calculation of the Gurney velocity of high explosives using the BKW thermochemical code, Journal of Energetic Materials, 33 (2015) 140–144.

[153] D. Frem, A reliable method for predicting the specific impulse of chemical propellants, Journal of Aerospace Technology and Management, 10 (2018).

[154] M. H. Keshavarz, Liquid Fuels As Jet Fuels and Propellants: A Review of Their Productions and Applications, Nova Science Publishers, Inc., New York, 2018.

[155] D. Frem, Simple correlations for the estimation of propellants specific impulse and the gurney velocity of high explosives, Combustion Science and Technology, 188 (2016) 77–81.

[156] B. T. Fedoroff, O. E. Sheffield, Encyclopedia of Explosives and Related Items, Part 2700, Picatinny Arsenal, Dower, NJ, 1974.

[157] L. C. Smith, On Brisance, and a Plate-denting Test for the Estimation of Detonation Pressure, Los Alamos Scientific Lab., N. Mex., 1963.

[158] T. R. Gibbs, A. Popolato, LASL Explosive Property Data, Univ. of California Press, Berkeley, 1980.

[159] M. Jafaria, M. Kamalvand, M. H. Keshavarz, A. Zamanib, H. Fazeli, A simple approach for prediction of the volume of explosion gases of energetic compounds, Indian Journal of Engineering & Materials Sciences, 22 (2015) 701–706.

[160] G. V. Belov, Thermodynamic analysis of combustion products at high temperature and pressure, Propellants, Explosives, Pyrotechnics, 23 (1998) 86–89.

[161] M. H. Keshavarz, A simple procedure for assessing the performance of liquid propellants, High Temperatures. High Pressures, 35 (2003) 587–592.

[162] M. H. Keshavarz, Prediction method for specific impulse used as performance quantity for explosives, Propellants, Explosives, Pyrotechnics, 33 (2008) 360–364.

[163] R. Gill, L. Asaoka, E. Baroody, On underwater detonations, 1. A new method for predicting the CJ detonation pressure of explosives, Journal of Energetic Materials, 5 (1987) 287–307.

[164] M. H. Keshavarz, H. R. Pouretedal, Predicting detonation velocity of ideal and less ideal explosives via specific impulse, Indian Journal of Engineering and Materials Sciences, 11 (2004) 429–432.

[165] M. H. Keshavarz, M. Ghorbanifaraz, H. Rahimi, M. Rahmani, Simple pathway to predict the power of high energy materials, Propellants, Explosives, Pyrotechnics, 36 (2011) 424–429.

[166] T. Boggs, D. Zurn, W. Strahle, J. Handley, T. Milkie, Mechanisms of Combustion, Naval Weapons Center, China Lake, 1978.

[167] S. Gordon, B. J. McBride, Computer Program for Calculation of Complex Chemical Equilibrium Compositions, Rocket Performance, Incident and Reflected Shocks, and Chapman–Jouguet Detonations, National Aeronautics and Space Adminstration (USA), 1976.

[168] M. H. Keshavarz, M. Ghorbanifaraz, H. Rahimi, M. Rahmani, A new approach to predict the strength of high energy materials, Journal of Hazardous Materials, 186 (2011) 175–181.

[169] B. T. Fedoroff, O. E. Sheffield, Encyclopedia of Explosives and Related Items, Part 2700, Picatinny Arsenal, Dower, NJ, 1972.

[170] M. Kamalvand, M. H. Keshavarz, M. Jafari, Prediction of the strength of energetic materials using the condensed and gas phase heats of formation, Propellants, Explosives, Pyrotechnics, 40 (2015) 551–557.

[171] M. Jafari, M. Kamalvand, M. H. Keshavarz, S. Farrashi, Assessment of the strength of energetic compounds through the Trauzl lead block expansions using their molecular structures, Zeitschrift für anorganische und allgemeine Chemie, 641 (2015) 2446–2451.

[172] S. M. Kaye, Encyclopedia of Explosives and Related Items, Part 2700, Defense Technical Information Center, New Jersey, 1978.

[173] M. H. Keshavarz, F. Seif, Improved approach to predict the power of energetic materials, Propellants, Explosives, Pyrotechnics, 38 (2013) 709–714.

[174] B. T. Fedoroff, O. E. Sheffield, E. F. Reese, O. E. Sheffield, G. D. Clift, C. G. Dunkle, H. Walter, D. C. Mclean, Encyclopedia of Explosives and Related Items, Part 2700, Picatinny Arsenal, Dower, NJ, 1960.

[175] M. H. Keshavarz, F. Seif, H. Soury, Prediction of the brisance of energetic materials, Propellants, Explosives, Pyrotechnics, 39 (2014) 284–288.

[176] B. T. Fedoroff, O. E. Sheffield, E. F. Reese, G. D. Clift, Encyclopedia of Explosives and Related Items, Part 2700, Picatinny Arsenal, Dower, NJ, 1962.

[177] H. Hornberg, F. Volk, The cylinder test in the context of physical detonation measurement methods, Propellants, Explosives, Pyrotechnics, 14 (1989) 199–211.

[178] D. Frem, Predicting the plate dent test output in order to assess the performance of condensed high explosives, Journal of Energetic Materials, 35 (2017) 20–28.

[179] D. Frem, The Use of the [H2O–CO2] arbitrary decomposition assumption to predict the performance of condensed high explosives, Combustion, Explosion, and Shock Waves, 54 (2018) 704–711.

[180] D. Frem, The specific impulse as an important parameter for predicting chemical high explosives performance, Zeitschrift für anorganische und allgemeine Chemie, 644 (2018) 235–240.

[181] D. Xiang, J. Rong, X. He, Z. Feng, Underwater explosion performance of RDX/AP-based aluminized explosives, Central European Journal of Energetic Materials, 14 (2017) 60–76.

[182] D. L. Xiang, J. L. Rong, J. Li, Effect of Al/O ratio on the detonation performance and underwater explosion of HMX-based aluminized explosives, Propellants, Explosives, Pyrotechnics, 39 (2014) 65–73.

[183] R. H. Cole, Underwater Explosions, Princeton University Press, Princeton, New Jersey, 1948.

[184] M. J. Lin, H. H. Ma, Z. W. Shen, X. Z. Wan, Effect of aluminum fiber content on the underwater explosion performance of RDX-based explosives, Propellants Explosives Pyrotechnics, 39 (2014) 230–235.

[185] M. M. Swisdak Jr, Explosion Effects and Properties. Part II. Explosion Effects in Water, U.S. Naval Surface Weapons Center White Oak Lab, Silver Spring, MD, 1978.

[186] E. Stromsoe, S. W. Eriksen, Performance of high explosives in underwater applications. Part 2: Aluminized explosives, Propellants, Explosives, Pyrotechnics, 15 (1990) 52–53.

[187] J. P. Lu, D. L. Kennedy, Modelling of PBX-115 Using Kinetic CHEETAH and the DYNA Codes, Weapons Systems Division Systems Sciences Laboratory, SA, Australia, 2003.

[188] M. Hagfors, J. Saavalainen, Underwater explosions Y particle size effect of Al powder to the energy content of an emulsion explosive, in: Annual Conference on Explosives and Blasting Technique, Orlando, Florida, USA 7–10 February 2010, International Society of Explosives Engineering, Cleveland, 2010.

[189] M. H. Keshavarz, V. Bagheri, A simple correlation for assessment of the shock wave energy in underwater detonation, Zeitschrift für anorganische und allgemeine Chemie, 645 (2019) 1146–1152.

[190] M. H. Keshavarz, V. Bagheri, S. Damiri, A simple method for reliable estimation of the bubble energy in the underwater explosion, Zeitschrift für anorganische und allgemeine Chemie, 645 (2019) 1402–1407.

[191] R. Simpson, P. Pagoria, A. Mitchell, C. Coon, Synthesis, properties and performance of the high explosive ANTA, Propellants, Explosives, Pyrotechnics, 19 (1994) 174–179.

[192] M. D. Coburn, Ammonium 2,4,5-trinitroimidazole, Panent No. 4,028,154, US Patents, 1977.

[193] H. Gao, J. M. Shreeve, Azole-based energetic salts, Chemical Reviews, 111 (2011) 7377–7436.

[194] D. L. Naud, M. A. Hiskey, J. F. Kramer, R. Bishop, H. Harry, S. F. Son, J. D. Sullivan, High-Nitrogen Explosives, in: 29th International Pyrotechnics Seminar, Westminster, USA, 2002.

[195] D. E. Chavez, M. A. Hiskey, D. L. Naud, Tetrazine explosives, Propellants, Explosives, Pyrotechnics, 29 (2004) 209–215.

[196] T. M. Klapötke, B. Krumm, F. X. Steemann, Preparation, characterization, and sensitivity data of some azidomethyl nitramines, Propellants, Explosives, Pyrotechnics, 34 (2009) 13–23.

[197] J. B. Pedley, Thermochemical Data of Organic Compounds, Springer Science & Business Media, 2012.

[198] M. B. Talawar, R. Sivabalan, S. N. Asthana, H. Singh, Novel ultrahigh-energy materials, Combustion, Explosion, and Shock Waves, 41 (2005) 264–277.

[199] V. I. Pepekin, Limiting detonation velocities and limiting propelling powers of organic explosives, Doklady. Physical Chemistry, 414 (2007) 159–161.

[200] D. L. Ornellas, The heat and products of detonation in a calorimeter of CNO, HNO, CHNF, CHNO, CHNOF, and CHNOSi explosives, Combustion and Flame, 23 (1974) 37–46.

[201] L. E. Fried, P. Clark Souers, BKWC: An empirical BKW parametrization based on cylinder test data, Propellants, Explosives, Pyrotechnics, 21 (1996) 215–223.

[202] H. Östmark, U. Bemme, H. Bergman, N-guanylurea-dinitramide: a new energetic material with low sensitivity for propellants and explosives applications, Thermochimica Acta, 384 (2002) 253–259.

[203] T. M. Klapötke, P. Mayer, J. Stierstorfer, J. J. Weigand, Bistetrazolylamines – synthesis and characterization, Journal of Materials Chemistry, 18 (2008) 5248–5258.

[204] P. P. Vadhe, R. B. Pawar, R. K. Sinha, S. N. Asthana, A. S. Rao, Cast aluminized explosives (Review), Combustion, Explosion, and Shock Waves, 44 (2008) 461–477.

[205] D. L. Ornellas, Calorimetric Determinations of the Heat and Products of Detonation for Explosives, Lawrence Livermore National Laboratory, 1982.

[206] T. Wei, J. Wu, W. Zhu, C. Zhang, H. Xiao, Characterization of nitrogen-bridged 1,2,4,5-tetrazine-, furazan-, and 1H-tetrazole-based polyheterocyclic compounds: heats of formation, thermal stability, and detonation properties, Journal of Molecular Modeling, 18 (2012) 3467–3479.

[207] E. D. Brown, An introduction to ProPEP, A propellant evaluation program for personal computers, Journal of Pyrotechnics, 1 (1995) 11–18.

[200] M. R. Manaa, I. F. W. Kuo, L. E. Fried, First principles high pressure unreacted equation of state and heat of formation of crystal 2,6-diamino-3, 5-dinitropyrazine-1-oxide (LLM-105), Journal of Chemical Physics, 141 (2014) 064702.

[209] A. M. Astachov, V. A. Revenko, E. S. Buka, Comparative characteristics of two isomeric explosives: 4-nitro-5-nitrimino-1H-1,2,4-triazole and 3-nitro-5-nitrimino-1,4H-1,2,4-triazole, in: Z. Zeman (Ed.) The 7th Seminar in "Research of Energetic Materials", New Trends in Research of Energetic Materials, Pardubice, Czech Republic, 2004, pp. 424–432.

[210] B. C. Tappan, S. F. Son, A. N. Ali, D. E. Chavez, M. A. Hiskey, Decomposition and performance of new high nitrogen propellants and explosives, in: Sixth International Symposium on Special Topics in Chemical Propulsion, Santiago, Chile, 2005.

Index

https://doi.org/10.1515/9783110677652-011

www.ingramcontent.com/pod-product-compliance
Lightning Source LLC
Chambersburg PA
CBHW081536220326
41598CB00036B/6458